Mammals, Amphibians, and Reptiles of Costa Rica

NUMBER SIXTY-SIX, *The Corrie Herring Hooks Series*

Foreword by WINNIE HALLWACHS

Mammals, Amphibians, and Reptiles of Costa Rica

A FIELD GUIDE

CARROL L. HENDERSON
with photographs by the author

Illustrations by STEVE ADAMS

UNIVERSITY OF TEXAS PRESS, *Austin*

The University of Texas Press wishes to acknowledge the generous financial support by the following foundations, individuals, and businesses that helped to underwrite the costs of producing this book:

The Dellwood Wildlife Foundation of Dellwood, Minnesota, in memory of wildlife conservationist and founder of the Dellwood Wildlife Foundation, Ramon D. (Ray) Whitney. Ray Whitney was instrumental in helping restore trumpeter swans to Minnesota, and he shared a love and appreciation for the diversity and abundance of wildlife in Costa Rica.

The Costa Rica–Minnesota Foundation of St. Paul, Minnesota, in support of cultural and natural resource initiatives fostering greater understanding, educational programs, and habitat protection for Costa Rica's wildlife.

The late Honorary Consul to Costa Rica from Minnesota and former CEO of the H. B. Fuller Company, Tony Andersen, who was a tireless promoter for cooperative projects of benefit to Costa Rica's culture and environment.

Karen Johnson, President of Preferred Adventures Ltd. of St. Paul, Minnesota, a business that specializes in ecotourism and natural history adventures for worldwide travelers. She has been especially active in promoting wildlife tourism in Costa Rica and other countries of Latin America.

Michael Kaye, President of Costa Rica Expeditions, San José, Costa Rica. Costa Rica Expeditions owns and manages Monteverde Lodge, Tortuguero Lodge, and Corcovado Lodge Tent Camp. This company has set high standards for protecting sensitive tropical habitats while accommodating the needs of nature tourism and adventure travelers in Costa Rica.

Dan Conaway, President of Elegant Adventures, Atlanta, Georgia. Elegant Adventures specializes in quality, customized tours to Latin American destinations, including Costa Rica. This company has served international travelers since its founding in 1986.

Printed in China
First edition, 2002
Field Guide to the Wildlife of Costa Rica, 2002
Birds of Costa Rica: A Field Guide, 2010
Butterflies, Moths, and Other Invertebrates of Costa Rica: A Field Guide, 2010
Mammals, Amphibians, and Reptiles of Costa Rica: A Field Guide, 2010

Requests for permission to reproduce material from this work should be sent to
Permissions
University of Texas Press
P.O. Box 7819
Austin, TX 78713-7819
www.utexas.edu/utpress

∞ The paper used in this book meets the minimum requirements of ANSI/NISO Z39.48-1992 (R1997) (Permanence of Paper).

LIBRARY OF CONGRESS
CATALOGING-IN-PUBLICATION DATA

Henderson, Carrol L.
Mammals, amphibians, and reptiles of Costa Rica : a field guide / Carrol L. Henderson ; photographs by Carrol L. Henderson ; illustrations by Steve Adams ; foreword by Winnie Hallwachs.
p. cm. — (The Corrie Herring Hooks series ; no. 66)
Includes bibliographical references and index.
ISBN 978-0-292-72274-3 (pbk. : alk. paper)
1. Mammals—Costa Rica—Guidebooks. 2. Amphibians—Costa Rica—Guidebooks. 3. Reptiles—Costa Rica—Guidebooks. 4. Costa Rica—Guidebooks. I. Title.
QL723.C8H46 2010
599.097286—dc22

2010001033

pg i: Red-eyed Tree Frog
pg ii: Young Three-toed Sloth
Facing pg: Backlit canal in Tortuguero National Park

To my wife, Ethelle,
and son and daughter-in-law, Craig and Reem,
with whom I share my love of Costa Rica,
and grandson Mazen Nathaniel,
and to
Drs. George Knaphus, James H. Jenkins, and Daniel H. Janzen,
my mentors.

CONTENTS

FOREWORD

The joy of living in a house in a Costa Rican dry forest conservation area is that anything might happen, anytime. The actors are mammals, reptiles, frogs, and insects that make use of the open house and hunt or shelter inside. Spectacular though they are, the birds stay out—except for their sounds—and the stage is left to other vertebrates. Much of the action happens toward dark or in the night. As the afternoon dims and the dusk cicadas sing, the Sac-Wing *Saccopteryx* bats leave the back wall and fly sweeping patrols back and forth, imperceptibly feeding on insects. The long-tongued, nectar-feeding *Glossophaga* bats stay indoors much later, but they are the aerial stars. During courtship, they fly around the rooms like breakneck stunt pilots, through all available three-dimensional space, up over walls and swooping down under the desk, across my lap and around again. In the still of the night the footfalls of the Virginia Opossum or the soft *Caluromys* quietly cross the metal roof. A *Liomys* mouse, the ancestor of the kangaroo rats in the desert of the United States, loads its cheek pouches to near bursting from a bag of rice, ferries it to store temporarily in a shoe or all the way into its burrow, and returns almost insatiably for more.

Dr. Winnie Hallwachs

In the daylight, the large *Trachycephalus* (*Phrynohyas*) tree frogs hide in pots or back into drains with rubbery contortions. The Sac-wing Bats have returned and are stretching and nursing and arguing and chattering audibly into the morning. Just outside, a vine moves strangely, rising, and comes into focus as one of the vine snakes, an *Oxybelis* or the big-headed *Imantodes* or the bright parrot snake *Leptophis mexicanus*. Don't leave to get a camera—look away and it will become invisible again. Lizards appear, from the small red-headed *Gonatodes* and taut metallic skink *Mabuya* on the shady walls to the large iguanalike ctenosaur males in the sunlit road and tree canopies.

These house experiences are close-up glimpses of what is happening over the grand sweep of the large conservation area—little windows into the richness and diversity of very different species going about their lives. Carrol and Ethelle Henderson have spent much of their adult lives bringing people to the natural world, or in Carrol's many books, the natural world to people. Here Carrol's dramatic new photographs and expanded text take you into that world again.

The tragic back story, though, is that wild Costa Rica is changing, as is all else in the natural world, severely affected directly and indirectly by human activity. Some of the generalist or human-commensal vertebrates have become common and widespread, but many of the specialist species are shrinking or on the way to local extinction, even in protected areas such as the conservation areas. Very many are becoming genetically stunted as the landscape changes and their habitats shrink and change from once large landmasses to archipelagos of little fragments, winking out one by one.

The Golden Toad is gone, and the great Pacific Leatherback that shakes the beach with its heaves is severely threatened. Behind these, a host of lesser-known species are shrinking or going extinct outside of human notice. Their absence ripples through the natural world, but humans are not aware. May this book help to take us toward an awakening.

WINNIE HALLWACHS
Sector Santa Rosa
Área de Conservación Guanacaste
Costa Rica
SEPTEMBER 2009

PREFACE

I grew up as a farm boy near Zearing in central Iowa, and most of my early travels were within twenty-five miles of our family farm. I had quite a provincial view of life and no concept of ecosystems, biological diversity, or tropical rainforests. I just knew that I loved wildlife. I had no idea that Costa Rica, a small country thousands of miles away in Central America, would later play such a dramatic role in shaping the direction of my personal and professional life.

An early and enthusiastic interest in nature led me to major in zoology and minor in botany at Iowa State University. After completing my bachelor's degree at ISU in 1968, I enrolled in graduate school at the University of Georgia, where I studied ecology, forest and wildlife management, journalism, and public relations. During my search for a thesis topic, Dr. James H. Jenkins directed me to an Organization for Tropical Studies (OTS) course in Costa Rica.

When I began my two-month OTS course in tropical grasslands agriculture in February of 1969, I had no idea it would be such a life-changing experience. Every day was an adventure! I tried to absorb all that I could about the land, the people, and the wildlife of Costa Rica. I quickly learned that this is not a country you can visit just once. By March I had already applied for another OTS course and was subsequently accepted. In June 1969, I drove from Georgia to Costa Rica with Dr. Jenkins for an OTS course in tropical ecology.

The author with an oropendola nest during an OTS course in Costa Rica, 1969.

The OTS faculty, recruited from educational institutions throughout North and Central America, included some of the most notable tropical biologists in the world. They inspired me with their knowledge and enthusiasm about tropical ecosystems. By the end of the tropical ecology course, I had fallen in love with the country, with its people, and with Ethelle González Álvarez, a student at the University of Costa Rica. I returned to Costa Rica a third time in 1969. Ethelle and I were married in December of that year and have now been married forty years. We have a son, Craig, who shares

Birders in forest.

our love and enthusiasm for his Tico heritage, along with his wife, Reem, and son, Mazen Nathaniel.

After returning to the University of Georgia, I wrote my master's thesis, "Fish and Wildlife Resources of Costa Rica, with Notes on Human Influences." The 340-page thesis analyzed human influences that were having significant positive or negative impacts on Costa Rica's wildlife. I also provided recommendations for changes in the game laws that would improve management of the country's wildlife.

During the four decades since my first visit to Costa Rica, I have returned forty-one times. Since 1987, our visits to the country have included leading wildlife tours. Ethelle and I have led twenty-six birding and wildlife tours to Costa Rica since 1987 in coordination with Preferred Adventures Ltd. of St. Paul. We continue to see new species on every visit—and every day is still an adventure!

Each year thousands of first-time tourists are experiencing the same sense of wonder about the country's rainforests and wildlife that I did in 1969, and Costa Rica has become one of the top nature tourism destinations in the world. This book is written to share my enthusiasm and knowledge about the country's wildlife with those tourists and with Costa Ricans who share our love of nature. It is written to answer questions about identification, distribution, natural history, and the incredible ecological adaptations of some of the most memorable and interesting mammals, amphibians, and reptiles in Costa Rica. It also provides the opportunity to recognize the people and conservation programs that have made Costa Rica a world leader in preserving its tropical forest and wildlife resources. This book will help you identify wildlife during your trip and relish your memories of this world-class wildlife destination after you return home and share the photos in the book with your friends and family.

ACKNOWLEDGMENTS

Writing this book has been a real labor of love. It represents the culmination of forty years of personal and professional relationships in Costa Rica. Special appreciation goes to my wife, Ethelle, and my son, Craig, who have traveled with me from Minnesota to Costa Rica many times and helped with everything from wildlife observations to editing and preparing the manuscript. In 1985, Karen Johnson, the owner of Preferred Adventures Ltd. in St. Paul, Minnesota, convinced us to try leading a birding trip to Costa Rica. We finally agreed and led our first trip in 1987. It was the beginning of a wonderful experience that has enabled us to meet many special people in our tour groups as well as Costa Rican tourism outfitters, guides, and ecolodge staffs.

Birding group in rainforest, 1987.

Michael Kaye, the owner of Costa Rica Expeditions, has been very supportive of this project and has coordinated our travel there. He facilitated travel to visit several sites for photography purposes, including Monteverde Lodge. Carlos Gómez Nieto is the extraordinary guide who has led all but one of our Costa Rican birding trips. Carlos is the premier birder in Costa Rica, and his vast knowledge of wildlife behavior and identification has helped us accumulate our wildlife records, which now exceed 27,000 observations. Carlos reviewed the manuscript for the book, and his wife, Vicky, also accompanied us on several wildlife outings. Manuel Salas and Marco Antonio "Niño" Morales have been the drivers for most of our trips and have been invaluable in spotting wildlife and in providing us with safe and memorable travel experiences.

Other people have helped with facilitating our travels and the collection of information and photos. They include Lisa and Kathy Erb at Rancho Naturalista; Don

Birders on a boardwalk, La Selva Biological Field Station.

Efraín Chacón, Rolando Chacón, and the rest of the Chacón family; Amos Bien of Rara Avis; Gail Hewson-Hull; Luis Diego Gómez; the late Dr. Alexander Skutch and Pamela Skutch at Los Cusingos; and the late Werner and Lily Hagnauer.

Biologists and scientists provided expertise on species identification and life history data, including Dr. Daniel Janzen, Dr. Graciela Candelas, the late Dr. Alexander Skutch, Brian Kubicki, Dr. Frank T. Hovore, Jorge Corrales of the Instituto Nacional de Biodiversidad (INBIO), and Dr. Jay M. Savage. Dr. Savage is an emeritus professor of biology at the University of Miami and adjunct professor of biology at San Diego State University. The author became acquainted with Dr. Savage during his work with the Organization for Tropical Studies in 1969. He continued to consult with Dr. Savage on questions relating to the identification of Costa Rica's herpetofauna during the preparation of the *Field Guide to the Wildlife of Costa Rica.*

The photos in this volume are by the author, except as follows: The photo of Dr. Dan Janzen teaching in 1967 was provided by the Organization for Tropical Studies. The Red Brocket Deer was photographed by Dr. Janzen. The beautiful photo of the Greater Fishing Bat is by Dr. J. Scott Altenbach, the rare Olingo photo is by Robert Djupstrom, the two White Tent Bat photos are by Joanna Eckles, and the Olive Ridley Turtle photos are by Pablo Vásquez Badilla. There are also twenty-two stunning photos of frogs and toads generously provided by Brian Kubicki.

And finally, special appreciation goes to all the Costa Rican travelers who have accompanied us on our trips and provided the companionship, sharp eyes, and friendships that have enriched our lives.

Mammals, Amphibians, and Reptiles of Costa Rica

The name “Costa Rica” means “Rich Coast.”

INTRODUCTION

Costa Rica! The name generates a sense of excitement and anticipation among international travelers. Among European explorers, the first recorded visitor was Christopher Columbus in 1502. On his fourth trip to the New World, Columbus landed where the port city of Limón is now located. The natives he encountered wore golden disks around their necks. He called this new place "Costa Rica," meaning "Rich Coast," because he thought the gold came from there. The gold had actually come from other countries and had been obtained as a trade item from native traders along the coast.

Spanish treasure seekers eventually discovered their error and went elsewhere in their quest for gold. The irony is that Christopher Columbus actually picked the perfect name for this country. The wealth overlooked by the Spaniards is the rich biological diversity that includes more than 505,000 species of plants and wildlife. That species richness is an incredible natural resource that sustains one of the most successful nature tourism industries in the Western Hemisphere. It also provides the basis for a newly evolving biodiversity industry of "chemical prospecting" among plants and creatures, in search of new foods and medicines for humans.

For such a small country, Costa Rica gets much well-deserved international attention and has become one of the most popular nature tourism destinations in the Americas. The lure is not "sun and sand" experiences at big hotels on the country's beaches; it is unspoiled nature in far-flung nooks and crannies of tropical wildlands that are accessible at rustic, locally owned nature lodges throughout the country.

It is now possible to immerse yourself in the biological wealth of tropical forests during a vacation in Costa Rica. During a two-week visit you may see three hundred to four hundred species of birds, mammals, reptiles, amphibians, butterflies, moths, and other invertebrates. Some vacations are planned for rest and relaxation, but who can do that in such a diverse country where there is so much nature to see and experience? In Costa Rica, every day is an adventure, and the marvelous diversity and abundance of wildlife create an enthusiasm for nature that many people have not experienced since childhood.

Every day is an adventure in Costa Rica.

The ease with which it is possible to travel to Costa Rica and enjoy wildlife in such a pristine setting makes a visitor think it has always been that way. It has not. The appealing travel and tourism conditions are the product of nearly five decades of social, educational, and cultural developments.

There was a time when Costa Rican wildlife was persecuted at every opportunity. Virtually every creature weighing over a pound was shot for its value as meat or for its hide. Wildlife was killed year-round from the time of settlement through the 1960s. Instead of acquiring souvenirs like T-shirts and postcards in those days, Costa Rican visitors in the 1960s found vendors selling boa constrictor hides, caiman-skin briefcases, stuffed caimans, skins of spotted cats, and sea turtle eggs.

HISTORICAL PERSPECTIVE

To appreciate the abundance of today's wildlife populations, it is necessary to understand the revolution in wildlife conservation and habitat preservation that has occurred since the 1960s. Dozens of dedicated biologists, politicians, and private citizens have contributed to Costa Rica's world leadership in tropical forest conservation, wildlife protection, and nature tourism over the past fifty-plus years. This process occurred in five phases: (1) Research, (2) Education, (3) Preservation, (4) Conservation, and (5) Nature Tourism.

Research

One of the earliest advances for Costa Rica's legacy of conservation was the development of research data on Costa Rica's plants and wildlife. Without such basic knowledge, there can be little appreciation, respect, or protection for wild species. In 1941, Dr. Alexander Skutch homesteaded property in the San Isidro del General Valley along the Río Peña Blanca. After settling there with his wife, Pamela, Dr. Skutch studied Costa Rica's birds for more than sixty years and continued to observe them and record their life history in his prolific writings until his passing in 2004.

In 1954, another biologist, Dr. Archie Carr, started epic research. Dr. Carr, from the University of Florida, began a lifelong commitment to the protection and management of the Green Turtle at Tortuguero. That effort continues to this day, thanks to his son, Dr. David Carr, and the work of the Caribbean Conservation Corporation, which was created in 1959.

Another significant development for Costa Rica's legacy of leadership in tropical research was the creation of the Tropical Science Center. It was founded in 1962 by Drs. Leslie R. Holdridge, Joseph A. Tosi, and Robert J. Hunter. These three scientists promoted research on tropical ecosystems, land use, and sustainable development. Dr. Gary Hartshorn later joined the staff to add more expertise in the development of tropical forest management strategies. The Tropical Science Center was instrumental in establishing La Selva Biological Field Station and the Monteverde Cloud Forest Reserve and in preserving Los Cusingos, the forest reserve formerly owned by Dr. Alexander Skutch. That reserve is now managed by the Tropical Science Center.

Another research catalyst for subsequent conservation and land protection was the creation of the Organization for Tropical Studies (OTS) in 1964. The OTS is a consortium of fifty-five universities and educational institutions throughout the

Americas. The OTS operates three tropical research field stations—located at La Selva, Palo Verde, and San Vito. Tropical biologists from throughout the world come to these field stations to pursue pioneering studies on taxonomy, ecology, and conservation of tropical ecosystems.

For many decades, people had believed it was necessary to eliminate tropical forests in the name of progress, to create croplands, pastures, and monocultures of exotic trees for the benefit of society. Tropical biologists of the OTS changed the way people viewed tropical forests and helped society realize the infinitely greater ecological, climatic, and economic benefits that can accrue from preserving and managing tropical forests as sustainable resources.

Education

In 1963 the National Science Foundation supported the Advanced Science Seminar in Tropical Biology, which was subsequently adapted by OTS. The OTS initiated a second part of its legacy with field courses in tropical ecology, forestry, agriculture, and land use for undergraduate and graduate students from throughout the Americas. Since its founding, the OTS has conducted more than 200 field courses for at least 3,600 students. For many of these students, including the author, the courses were life-changing experiences. The faculty who taught these courses were some of the most prominent ecologists in the world, including, among others, Drs. Dan Janzen, Mildred Mathias, Carl Rettenmeyer, Frank Barnwell, Rafael Lucas Rodríguez Caballero, Gordon Orions, Roy McDiarmid, Larry Wolf, and Rex Daubenmire.

Dr. Dan Janzen teaching an OTS course in 1967. Photo provided courtesy of the Organization for Tropical Studies.

Another significant source of tropical education and research has been the Tropical Agricultural Center for Research and Education (Centro Agronómico Tropical de Investigación y Enseñanza; CATIE). This center was created in 1942 at Turrialba and was originally known as the Interamerican Institute of Agricultural Science (Instituto Interamericano de Ciencias Agrícolas; IICA). Graduate students come from all over Latin America to study agriculture, forestry, and wildlife management there.

Preservation

By the 1960s, about 50 percent of Costa Rica's forests had been cut, and the clearing continued. It became apparent that national programs for protection of the remaining forests and wildlife would be necessary if they were to be preserved into the next century.

The first wildlife conservation law was decreed on July 20, 1961, and was updated with bylaws on June 7, 1965. These laws and regulations provided for the creation and enforcement of game laws, the establishment of wildlife refuges, the prohibition of commercial sale of wildlife products, the issuance of hunting and fishing licenses, the establishment of fines for violations, and the creation of restrictions on the export and import of

Logging in Costa Rica, 1969

wildlife. Complete protection was given to tapirs, manatees, White-tailed Deer does accompanied by fawns, and Resplendent Quetzals. The laws, however, were not enforced.

In 1968, a Costa Rican graduate student, Mario Boza, was inspired by a visit to the Great Smoky Mountains National Park. In 1969 a Forestry Protection Law allowed national parks to be established, and Mario Boza was designated as the only employee of the new National Parks Department. He wrote a master plan for the newly designated Poás Volcano National Park as his master's thesis subject.

In 1970, wildlife laws were still being ignored by poachers, and wildlife continued to disappear. President "Don Pepe" Figueres visited Dr. Archie Carr and graduate student David Ehrenfeld to see the Green Turtle nesting beaches at Tortuguero. He was considering a proposal to protect the area as a national park. The following account was later written by Dr. David Ehrenfeld (1989):

It was Don Pepe's first visit to the legendary Tortuguero—we had been watching a Green Turtle nest, also a first for him. El Presidente, a short, Napoleonic man with boundless energy, was enjoying himself enormously. Both he and Archie were truly charismatic people, and they liked and respected one another. The rest of us went along quietly, enjoying the show. As we walked up the beach towards the boca, where the Río Tortuguero meets the sea, Don Pepe questioned Dr. Carr about the Green Turtles and their need for conservation. How important was it to make Tortuguero a sanctuary? Just then, a flashlight picked out a strange sight up ahead.

A turtle was on the beach, near the waterline, trailing something. And behind her was a line of eggs which, for some reason, she was depositing on the bare, unprotected sand. We hurried to see what the problem was.

When we got close, it was all too apparent. The entire undershell of the turtle had been cut away by poachers who were after calipee, or cartilage, to dry and sell to the European turtle soup manufacturers. Not interested in the meat or eggs, they had evidently then flipped her back on her belly for sport, to see where she would crawl. What she was trailing was her intestines. The poachers had probably been frightened away by our lights only minutes before.

Dr. Carr, who knew sea turtles better than any human being on earth and who had devoted much of his life to their protection, said nothing. He looked at Don Pepe, and so did I. It was a moment of revelation. Don Pepe was very, very angry, trembling with rage. This was his country, his place. He had risked his life for it fighting in the Cerro de la Muerte.

> The turtles were part of this place, even part of its name: Tortuguero; . . . She was home, laying her eggs for the last time.

Don Pepe realized that the ancient turtles, as well as the Costa Rican people, needed a safe place to live and raise their young. The poaching had to end. He declared Tortuguero National Park by executive decree in 1970. The tragic poaching incident with the nesting Green Turtle was probably the pivotal incident that catalyzed the national parks movement in Costa Rica. Mario Boza served under President Figueres as the National Park Service director from 1970 to 1974. By the end of 1974, the service had grown to an organization of 100 employees with an annual budget of $600,000. A total of 2.5 percent of the country was designated as national parks and reserves.

Private preservation efforts also began in the 1970s. Scientists George and Harriet Powell and Monteverde resident Wilford Guindon created the 810-acre Monteverde Cloud Forest Reserve. They brought in the Tropical Science Center to own and manage the preserve, which now totals 27,428 acres. The Monteverde Conservation League was subsequently formed to help manage and carry out conservation projects and land acquisition.

In 1984, Dr. Dan Janzen brought more international recognition to Costa Rica when he received the Crafoord Prize in Coevolutionary Ecology from the Swedish Royal Academy of Sciences. This is the ecologist's equivalent of the Nobel Prize. Dr. Janzen received the prize for his pioneering research on entomology and ecology of tropical dry forests. This focused attention on the need for preserving tropical dry forests in the Guanacaste Conservation Area (http://www.acguanacaste.ac.cr).

Oscar Arias was elected president in 1986. He created the Ministry of Energy, Mining, and Natural Resources (MIRENEM) by merging the national land

Rainbow over the Monteverde Cloud Forest Reserve

Dan Janzen with parataxonomist Osvaldo Espinoza of the Guanacaste Conservation area

management departments to make them more efficient in managing the nation's natural resources. That agency is now referred to as the Ministry of Environment and Energy (Ministerio del Ambiente y Energía; MINAE).

Since then, Costa Rica's national system of parks and reserves has continued to grow and mature. It now consists of 161 protected areas, including twenty-five national parks. Those areas total 3,221,635 acres—almost 26 percent of the country's land area. More information on the national park system can be found at www.costarica-nationalparks.com.

Conservation

As the national park system grew and encompassed more life zones, it became clear to ecological visionaries like Dr. Dan Janzen and Dr. Rodrigo Gámez Lobo that they had an opportunity to take another bold step that would place them in a world-leadership role for conservation of biological diversity and creation of economic benefits to society from that biological diversity. They created the National Institute of Biodiversity (Instituto Nacional de Biodiversidad; INBIO). Dr. Gámez Lobo became the first director of INBIO and continues his leadership as president of INBIO. The ambitious goal of this institute was to collect, identify, and catalog all of the living species in Costa Rica. Estimated at 505,000 species, this figure includes 882 birds, 236 mammals, 228 reptiles, 178 amphibians, 360,000 insects, and 10,000 plants. This represents about 5 percent of the world's species. So far, about 90,000 of those species have been described, and INBIO's collections include 3.5 million specimens.

Following the creation of INBIO, MIRENEM developed a national system of "conservation areas" in 1990. This is referred to as SINAC (Sistema Nacional de Areas de Conservación). Eleven conservation areas were established. Personnel in the fields of wildlife, forestry, parks, and agriculture teamed up to manage the national parks and wildlands in each conservation area. Their goal is the conservation of Costa Rica's biodiversity for nondestructive use by Costa Ricans and the world populace. This national scale of ecosystem-based management predated efforts in more "developed" countries by years.

Nature Tourism

Beginning in the mid-1980s, and concurrent with the conservation phase, the value of Costa Rica's national parks (NPs), national wildlife refuges (NWRs), and biological reserves (BRs) was reaffirmed in another way: as a resource for nature tourism. Nature tourism is motivated by the desire to experience unspoiled nature: to see, enjoy, experience, or photograph scenery, natural communities, wildlife, and native plants. The first rule

Birding guide Carlos Gómez Nieto

of nature tourism is that wildlife is worth more alive—in the wild—than dead. It has become a great incentive to protect wildlife from poachers and to conserve the forests as habitat for the wildlife.

Nature tourism provided new employment opportunities for Costa Ricans as travel agency personnel, outfitters, nature lodge owners and staff, drivers, and naturalist guides. The best naturalist guides can identify birds, mammals, reptiles, amphibians, flowers, butterflies, and trees. These dedicated guides share an infectious enthusiasm for the country as they help visitors experience hundreds of species during a visit. Guides like Carlos Gómez Nieto have seen more than 730 of the country's bird species and can identify most of them by sight and sound.

Rainforests—especially the international loss of rainforests—received a great deal of publicity in the 1980s. Costa Rica's national parks provided an opportunity to attract tourists to experience the mystique and beauty of those forests. Improved road systems and small airstrips throughout the country provided access to those parks and private reserves in the rainforests. Enterprising outfitters recognized the opportunity to establish locally owned and managed nature tourism lodges to cater to this new breed of international tourists.

One of the best-known pioneers in nature-based tourism is Michael Kaye. Originally from New York, he was a white-water rafting outfitter in the Grand Canyon before he founded Costa Rica Expeditions in 1985. In Costa Rica he provides tourists with the opportunity for adventure tourism, including white-water rafting and wildlife viewing. Kaye built three lodges—Tortuga, Monteverde, and Corcovado—and he provided ecologically based innovations and adaptations at these facilities that minimized their impact on the environment and sensitized visitors to the importance and vulnerability of the forests where these lodges were located. Kaye believes that tourists respond to world-class facilities and services that are provided by local ownership and management of smaller, dispersed lodging facilities. Costa Rica's tourism forte is that it is one of the best rainforest destinations in the Americas because it is safe and easily accessible; the attraction is not the beaches where huge hotels are owned by corporations from other countries.

There are now dozens of other locally owned nature-based lodges throughout the country. John Aspinall founded Costa Rica Sun Tours and built the Arenal Observatory Lodge. His brother Peter founded Tiskita Jungle Lodge. Don Perry initiated the Rainforest Aerial Tram facility. John and Kathleen Erb founded Rancho Naturalista and Tárcol Lodge. John and Karen Lewis founded Lapa Ríos. Other local nature lodges are Selva Verde, El Gavilán, Hacienda Solimar, La Pacífica, Caño Negro Lodge, La Ensenada, Villa Lapas, Rancho Casa Grande, Rara Avis, Savegre Mountain Lodge, Bosque de Paz,

Drake Bay Wilderness Resort, and El Pizote Lodge. OTS research facilities like La Selva and San Vito also provide accommodations for nature tourists.

International connections also benefited Costa Rica's nature tourism. In 1989, Preferred Adventures Ltd. was founded by Karen Johnson in St. Paul, Minnesota, with special emphasis on Costa Rican tourism. Through the efforts of Karen Johnson and the late Tony Andersen, Chairman of the Board and former CEO of the H. B. Fuller Company and Honorary Consul to Costa Rica from Minnesota, these connections resulted in the creation of the Costa Rica–Minnesota Foundation, which has promoted cultural, medical, and conservation projects in Costa Rica.

As nature tourism lodges proliferated after the mid-1980s, the number of tourists arriving in Costa Rica grew steadily. In 1988, about 330,000 tourists came, and over the next eleven years the number increased to 1,027,000 people per year. By 2006, the number of tourists had risen to 1,716,000. From 1988 to 2006, the annual number of national park visitors increased from about 865,600 to 1,205,100, and the number of tour operators increased from 58 to 325. All of this is happening in an area only one-fourth the size of Minnesota.

Nature tourism has turned heads throughout Costa Rica and Latin America because of the amount of income it has generated. In 1991, tourism contributed $330 million to the Costa Rican national economy. By 1999 that figure had increased to $940 million and exceeded the amount generated by exports of coffee ($408 million) and bananas ($566 million)! By 2006, tourism income exceeded one billion dollars: $1,620,800,000. In addition, nature tourism created more than 140,000 jobs.

Ctenosaur sharing the beach with tourists, Tamarindo.

The best thing about nature tourism is that when it is practiced ethically and in balance with the environment, it is a sustainable natural resource use that diversifies the economic base of the country and makes the value of the national parks and wildlife resources obvious to the citizenry.

Costa Ricans now realize that a significant part of the economic health and prosperity of their country is tied to the health and prosperity of their national parks, forests, and wildlife and to the future of the country's macaws, quetzals, tepescuintles, jaguars, and Green Turtles. Don Pepe Figueres was right. If you make the world a safe place for Green Turtles and other wildlife, it becomes a better place for people, too.

GEOGRAPHY

Costa Rica, a Central American country between Panama and Nicaragua, is shown in Figure 1. Considering its relatively small size, 19,653 square miles, Costa Rica has an exceptionally high diversity of plants and wildlife, more than half a million species. This is explained in part by the fascinating geological history of the region.

The geological history that led to the creation of Costa Rica goes back about 200 million years to the Triassic Period, when much of the earth's landmass was composed of a supercontinent called Pangaea. The

Figure 1. Location of Costa Rica in Central America.

supercontinent began to separate through continental drift, portrayed in Figure 2, which is the process by which the earth's landmasses essentially float on the molten core of the earth, drift among the oceans, and occasionally separate or merge. Pangaea eventually separated into two supercontinents. The northern supercontinent, called Laurasia, later became North America, Asia, and Europe. The southern portion, called Gondwanaland, drifted apart and later became South America, Africa, southern Asia, and Australia.

About 130 million years ago the western portion of Gondwanaland began to separate into South America and Africa. Concurrently, the North American landmass drifted westward from the European landmass. Both North America and South America drifted westward, but they were still separate. By the Pliocene Period, about three to four million years ago,

Highway map of Costa Rica. Source: U.S. State Department.

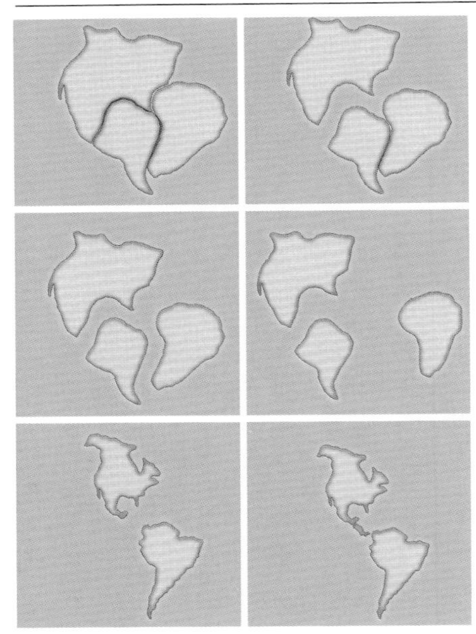

Figure 2. Stages in the process of continental drift that led to the creation of Costa Rica.

North America and South America were aligned from north to south, but a gap in the ocean floor between the two continents existed where southern Nicaragua, Costa Rica, and Panama are today.

About three million years ago an undersea plate of the earth's crust, called a tectonic plate, began moving north and eastward in the Pacific Ocean into the area between North and South America. This particular tectonic plate, the Cocos Plate, pushed onto the Caribbean Plate and rose above sea level to create the land bridge that now connects North and South America. That bridge became southern Nicaragua, Costa Rica, and the central and western portions of Panama.

BIOGEOGRAPHY

Biogeography is the relationship between the geography of a region and the long-term distribution and dispersal patterns of its plants and wildlife. The geological history of Costa Rica, Nicaragua, and Panama created a situation in which they became a land bridge between two continents. Plants and wildlife have been dispersing across that bridge for the last three million years; as a result, Costa Rica became a biological mixing bowl of species from both continents. Those dispersal patterns are shown in Figure 3.

Temperate-climate plants that have dispersed southward from North America include alders (*Alnus*), oaks (*Quercus*), walnuts (*Juglans*), magnolias (*Magnolia*), blueberries (*Vaccinium*), the Indian paintbrush (*Castilleja*), and the mistletoe (*Gaiadendron*). Most dispersal appears to have occurred during cooler glacial periods. As the climate became warmer, these northern-origin plants became biologically stranded on the mountains, where the climate was cooler.

Mammals dispersing from North America across the land bridge included coyotes, tapirs, deer, jaguars, squirrels, and bears. Birds that dispersed from North America to Central and South America included wrens, thrushes, sparrows, woodpeckers, and common dippers.

Plants that dispersed from South

Topographical relief map of Costa Rica.

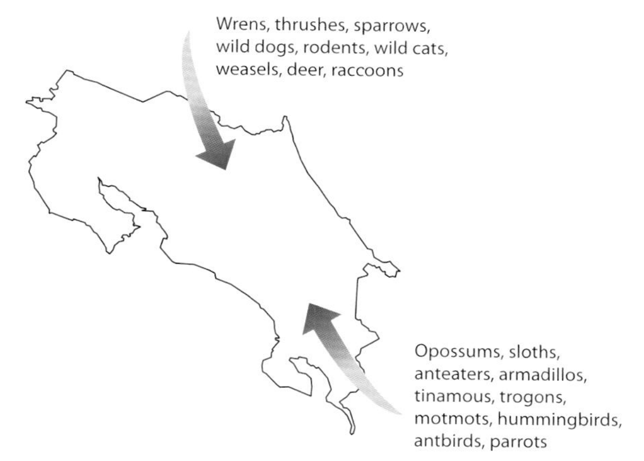

Figure 3. Costa Rica became a land bridge that facilitated the dispersal of wildlife from both North and South America.

America toward the north included tree ferns, cycads, heliconias, bromeliads, orchids, Poor-Man's Umbrella (*Gunnera*), *Puja*, and *Espeletia*. Mammals that dispersed northward from South America through Costa Rica are opossums, armadillos, porcupines, sloths, monkeys, anteaters, agoutis, and tepescuintles. Some species expanded through Central America and Mexico to the United States. Birds that dispersed from South America to Costa Rica and beyond include tinamous, hummingbirds, motmots, trogons, spinetails, flowerpiercers, antbirds, parrots, and woodcreepers.

ENDEMIC SPECIES

An endemic species is one found in one country or region and nowhere else in the world. Costa Rica has three zones of endemism, in which unique species and subspecies are found. These zones occur because geographic barriers created by mountains, arid zones, or oceans have isolated a particular population from the rest of its species, and eventually they evolve into separate species through natural selection.

Endemic Wildlife of the Highlands

The mountain ranges and volcanoes of Costa Rica and western Panama constitute an important zone of endemism for both plants and animals. The concept of endemism is often applied to species in a single country, but in the case of the Costa Rican highlands, where the Talamanca Mountains are contiguous with those of western Panama, the area of endemism crosses the border. For the purposes of this book, it is considered a single endemic zone. Many highland species have evolved into separate species or subspecies because they were reproductively isolated from other populations of the same species or similar species in the mountains of Guatemala and southern Mexico and in the mountains of eastern Panama and Colombia. This highland endemic zone is portrayed in Figure 4.

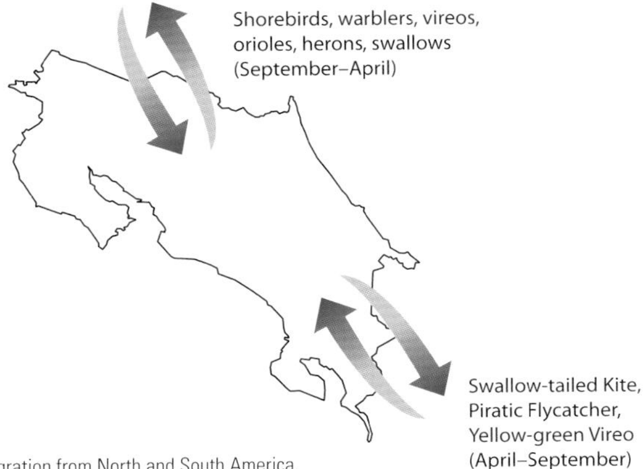

Figure 4. Patterns of bird migration from North and South America.

An impressive forty-seven birds are endemic to the mountains and foothills of Costa Rica and western Panama, and there are more than fifty subspecies of birds endemic to that region. The only regional endemic mammal in the mountains of Costa Rica and western Panama is the Poás Mountain Squirrel (*Syntheosciurus brochus*).

Endemic frogs of the highlands of Costa Rica and western Panama include *Craugastor podiciferus* and *Isthmohyla lancasteri*.

Endemic Species of the Southern Pacific Lowlands

Costa Rica's mountain ranges serve as a giant barrier that separates moist and wet lowland rainforest species that originally dispersed from South America to both the Caribbean lowlands and the southern Pacific lowlands. The mountains have caused reproductive isolation between populations of species that occurred in both areas. Over geologic time, the species diverged into separate species. This has

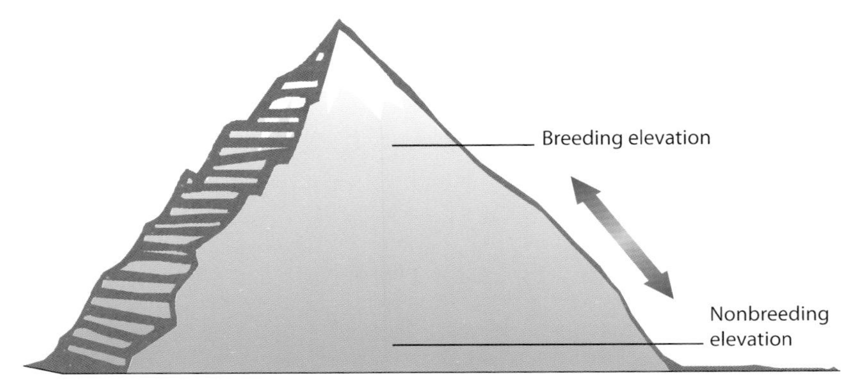

Figure 5. Some tropical insects and birds, like the Three-wattled Bellbird, carry out elevational migrations between breeding seasons and nonbreeding seasons.

contributed to a second zone of endemism in Costa Rica, the southern Pacific lowlands. There are several interesting pairs of species that have a common ancestor but have been separated from each other by the mountains between the Caribbean lowlands and the southern Pacific lowlands, as shown in Figure 5. Since the range of one species does not overlap the range of the other species in the pair, these are referred to as allopatric species.

This divergence of two species from a common ancestor is a continuing process, as evidenced by the recent decision by taxonomists to split the Scarlet-rumped Tanager into two species, Cherrie's Tanager in the Pacific lowlands and Passerini's Tanager in the Caribbean lowlands. The males are identical, but the females are distinctive. Other birds designated as endemic subspecies include the Masked Yellowthroat (Chiriquí race) and the Variable Seedeater (Pacific race). One day they may eventually become different enough from the Caribbean subspecies to be designated as new species. A pair of apparently allopatric reptile species includes the Central American Bushmaster in the Caribbean lowlands and the Black-headed Bushmaster in the southwest lowlands.

Additional endemic species of the southern Pacific lowlands that do not have a corresponding closely related species in the Caribbean lowlands include Baird's Trogon, Black-cheeked Ant-Tanager, Granular Poison Dart Frog, Golfo Dulce Anole, Mangrove Hummingbird, *Heliconius ismenius clarescens*, and Red-backed Squirrel Monkey.

Endemic Species of Cocos Island

A third zone of endemism is Cocos Island. This island is 600 miles out in the Pacific,

Figure 6. Highland zone of endemic species in Costa Rica and western Panama.

and three endemic birds have evolved there: Cocos Finch, Cocos Cuckoo, and Cocos Flycatcher. Cocos Island is an extension of the Galápagos Island archipelago but it is owned by Costa Rica. There are thirteen finches on the Galápagos Islands commonly referred to as Darwin's finches. The Cocos Finch is actually the fourteenth Darwin's finch. Undoubtedly, a variety of invertebrates are endemic to this island also.

MAJOR BIOLOGICAL ZONES

The most detailed and traditional classification of the habitats in Costa Rica includes twelve "life zones," as described by the late Dr. Leslie Holdridge of the Tropical Science Center. Those life zones are based on average annual precipitation, average annual temperature, and evapotranspiration potential. Evapotranspiration potential involves the relative amount of humidity or aridity of a region.

For tourism planning purposes, that

classification system has been simplified in this book from twelve to six biological zones. These zones, the first five of which are shown in Figure 8 on page 20, coincide with the distribution of many Costa Rican wildlife species and are designed for trip planning by wildlife tourists. The sixth zone consists of the entire coastline of both the Pacific and the Caribbean coasts. A good trip itinerary should include at least three biological zones in addition to the Central Plateau.

Tropical Dry Forest

The tropical dry forest in northwestern Costa Rica is a lowland region that generally coincides with the boundaries of Guanacaste Province. It extends eastward to the Cordilleras of Guanacaste and Tilarán, southeast to Carara NP, and north to the Nicaragua border. This zone extends from sea level to approximately 2,000 feet in elevation.

This region is characterized by a pronounced dry season from December through March. The deciduous trees include many plants that lose their leaves during the dry season and flower during that leafless period. Common trees are the bullhorn acacia (*Acacia*), *Tabebuia*, strangler fig (*Ficus*), *Guazuma*, kapok (*Ceiba*), *Bombacopsis*, buttercup tree (*Cochlospermum*), *Anacardium*, and the national tree of Costa Rica, the Guanacaste tree (*Enterolobium cyclocarpum*). The tallest trees approach 100 feet in height. Rainfall ranges from 40 to 80 inches per year.

Epiphytes are not a major component of the dry forest canopy, as they are in the moist and wet forests. However, some trees are thickly covered with vines like

Tropical dry forest in Guanacaste.

monkey vine (*Bauhinia*) and *Combretum*. The ease with which this forest can be burned and cleared for agricultural purposes has made the tropical dry forest the most endangered habitat in the country.

An important habitat within the dry forest consists of the riparian forests along the rivers, also called gallery forests. They maintain more persistent foliage during the dry season.

Wetlands, estuaries, islands, and backwaters of this region's rivers are also a major habitat for wetland wildlife. Especially important are lands along the Río Tempisque, its tributaries, and the wetlands of Palo Verde NP. Among reptiles characteristic of the dry forest are *Cnemidophorus deppii* and *Crotalus durissus*. Important examples of tropical dry forest habitat are preserved in Guanacaste Province; Santa Rosa, Las Baulas, and Palo Verde NPs; and Lomas Barbudal BR. The southeastern limit of this region is at Carara NP, which has a combination of wildlife characteristic of both the dry forest and the southern Pacific lowlands.

Southern Pacific Lowlands

The southern Pacific lowlands include the moist and wet forested region from Carara NP through the General Valley, Osa Peninsula, and Golfo Dulce lowlands to the Panama border and inland to the premontane forest zone at San Vito las Cruces.

The moist and wet forests of this region receive 80 to 200 inches of rainfall per year, with a more pronounced dry season from December through March than occurs in the Caribbean lowlands. These forests have fewer epiphytes than are found in Caribbean lowland forests. The tallest trees exceed 150 feet in height.

Among tree species are the kapok

Southern Pacific lowland forest, Manuel Antonio National Park.

(*Ceiba*), *Anacardium*, strangler fig (*Ficus*), wild almond (*Terminalia*), purpleheart (*Peltogyne purpurea*), *Carapa*, buttercup tree (*Cochlospermum vitifolium*), *Virola*, balsa (*Ochroma*), milk tree (*Brosimum*), *Raphia*, garlic tree (*Caryocar costaricense*), and *Hura*. Understory plants include species like bullhorn acacia (*Acacia*), walking palm (*Socratea*), *Bactris*, and *Heliconia*. Most trees maintain their foliage throughout the year.

Much of this region has been converted to pastureland and plantations of pineapple, coconut, and African oil palm. Among the most significant reserves remaining in natural habitat are Carara, Manuel Antonio, and Corcovado NPs. Corcovado NP is one of the finest examples of lowland wet forest in Central America, and it has excellent populations of wildlife species that are rare in other regions, such as Scarlet Macaws, jaguars, tapirs, and White-lipped Peccaries. Additional private reserves include one near San Isidro del General at Los Cusingos, the former home of Dr. Alexander and Pamela Skutch. It is now managed by the Tropical Science Center. The Wilson Botanical Garden at San Vito is an excellent example of premontane wet

Central Plateau overlooking San José and suburbs.

forest and is owned and operated by the Organization for Tropical Studies. The southern Pacific lowland area is of biological interest because it is the northernmost range limit for some South American species, for example, the Smooth-billed Ani, Masked Yellowthroat, Thick-billed Euphonia, Pearl Kite, Streaked Saltator, and Red-backed Squirrel Monkey. The Pearl Kite, Southern Lapwing, and Crested Oropendola are new arrivals and have all dispersed from Panama to this region since 1999. There have been unconfirmed records of the Western Night Monkey (*Aotus lemurinus*) in Costa Rica near the Panama border. The species occurs in northwestern Panama. Geoffroy's Tamarin (*Saguinus geoffroyi*) and the Bush Dog (*Speothos venaticus*) occur in Panama but have not yet been documented in Costa Rica.

Premontane (middle-elevation) sites like the Wilson Botanical Garden at San Vito are included in this biological region because many of the species typical of this region are found up to about 4,000 feet along the western slopes of the Talamanca Mountains. Premontane forests, like those preserved at the Wilson Botanical Garden, are the second most endangered life zone in Costa Rica, after tropical dry forests.

Central Plateau (Central Valley)

The Central Plateau contains the human population center of Costa Rica. The capital, San José, and adjoining suburbs are located in this relatively flat plateau at an elevation of approximately 3,900 feet. It is bordered on the north and east by major volcanoes of the Central Cordillera: Barva, Irazú, Poás, and Turrialba. To the south is the northern end of the Talamanca Mountains.

Rainfall ranges from 40 to 80 inches per year, and the original life zone in this area was premontane moist forest, but that forest has been largely cleared. The climate of the region, about 68 degrees Fahrenheit year-round, made it ideal for human settlement, and the rich volcanic soils made it an excellent region for growing coffee and sugarcane. The region is also important for production of fruits, vegetables, and horticultural export products like ferns and flowers.

Although premontane moist forests of the Central Plateau are largely gone, extensive plantings of shrubs, flowers, and fruiting and flowering trees throughout the San José area have made it ideal for adaptable wildlife species. Shade coffee plantations are preferred habitats for songbirds, including Neotropical migrants. Living fence posts of *Erythrina* and *Tabebuia* are excellent sources of nectar for birds and butterflies. Private gardens abound with butterflies and Rufous-tailed Hummingbirds. Remaining natural places, like the grounds of the Parque Bolívar Zoo in San José, host many wild, free-living butterflies, songbirds, Red-tailed Squirrels, and Long-tailed Weasels.

Among wildlife commonly encountered in backyards, woodlots, and open spaces of the Central Plateau and San José are the Clay-colored Thrush, Rufous-tailed Hummingbird, Blue-crowned Motmot, Tennessee Warbler, Blue-gray and Summer Tanagers, Rufous-collared Sparrow, Great-tailed Grackle, Broad-winged Hawk, Red-billed Pigeon, Crimson-fronted Parakeet, Groove-billed Ani, Hoffmann's Woodpecker, Tropical Kingbird, Social Flycatcher, Great Kiskadee, Blue-and-white Swallow, Brown Jay, and Baltimore Oriole. Among the mammals of the Central Plateau are numerous bats, the Variegated Squirrel, the Nine-banded Armadillo, Central American agoutis, Dice's Cottontail Rabbit, and the White-nosed Coati.

Caribbean Lowlands

The Caribbean lowlands include moist and wet lowland forests from the Caribbean coast westward to the foothills of Costa Rica's mountains. The Caribbean lowland fauna extends from the Río Frío and Los Chiles area southeastward to Cahuita and the Panama border. For the purposes of this book, the region extends from sea level to the upper limit of the tropical zone at about 2,000 feet elevation. The premontane forest, at least up to about 3,200 feet, also contains many lowland species. This region receives 80 to 200 inches of rainfall annually.

The trees grow to a height of over 150 feet. This is an evergreen forest that receives precipitation throughout the year and does not have a pronounced dry season like the moist and wet forests of the southern Pacific lowlands. Trees include coconut palms (*Cocos*), raffia palms (*Raphia*), *Carapa, Pentaclethra,* kapok (*Ceiba*), swamp almond (*Dipteryx panamensis*), *Alchornea,* walking palm (*Socratea*), and *Pterocarpus.* Tree branches have many epiphytes, such as bromeliads, philodendrons, and orchids. The complexity of the forest canopy contributes to a high diversity of plant and animal species in the treetops. Plants of the understory and forest edge include passionflower (*Passiflora*), *Hamelia, Heliconia,* palms, *Costus,* and *Canna.*

Much of this region has been cleared

Caribbean lowland wet forest, Tortuguero National Park.

and settled for production of cattle and bananas. Remaining forest reserves include Tortuguero and Cahuita NPs, Gandoca-Manzanillo NWR, Hitoy-Cerere BR, lower elevations of La Amistad and Braulio Carrillo NPs, and Caño Negro NWR. Tortuguero NP is one of the most extensive reserves and one of the best remaining examples of rainforest in Central America. Canals at Tortuguero and open water of the Río Frío and at Caño Negro provide excellent opportunities for viewing wildlife from boats. The grounds of Rara Avis also provide an excellent protected reserve at the upper elevational limit of this biological zone. La Selva Biological Field Station, owned and managed by the OTS, has an exceptional boardwalk and trail system that allows easy viewing of rainforests and rainforest wildlife.

Lower levels of Braulio Carrillo NP offer excellent examples of moist and wet lowland forest.

The Caribbean lowlands are significant as an excellent example of tropical habitat that supports classic rainforest species in all their complex diversity, beauty, and abundance: Great Green Macaws, Chestnut-mandibled and Keel-billed Toucans, trogons, jacamars, manakins, antbirds, parrots, tinamous, spinetails, Collared Peccaries, tapirs, howler monkeys, spider monkeys, White-faced Monkeys, Two- and Three-toed Sloths, bats, morpho and owl butterflies, Strawberry Poison Dart Frogs, and Red-eyed Tree Frogs. Among reptiles characteristic of the Caribbean lowlands (and also of the southwestern Pacific lowlands) are the Green Iguana, *Norops lemurinus*, *Ameiva festiva*, *Thecadactylus rapicaudus*, *Leptophis ahaetulla*, and the Spectacled Caiman.

At a time of increasing concern about the status of the world's amphibians and the effects of global warming, disease, and parasites on amphibian populations throughout the Americas, Costa Rica's Caribbean lowlands still appear to serve as a biologically rich reserve with an abundance of frog species and populations. The pioneering work of Brian Kubicki in the Caribbean foothills near Siquirres has helped identify endemic or restricted populations of such frog species as *Incilius melanochlorus*, *Cochranella spinosa*, *Craugastor bransfordii*, *Oophaga pumilio*, *Cruziohyla calcarifer*, and *Hyloscirtus palmeri*.

Highlands

The highland biological zone comprises Costa Rica's four mountain ranges. This zone includes lower montane, montane, and subalpine elevations generally above 4,200 to 4,500 feet in elevation.

Five volcanoes near the Nicaragua border form the Cordillera of Guanacaste: Orosí, Rincón de la Vieja, Santa María, Miravalles, and Tenorio.

The second group of mountains is the Cordillera of Tilarán. It includes the still-active Arenal volcano, which exploded in 1968, and mountains that are part of the Monteverde Cloud Forest.

Third is the Central Cordillera, which includes three large volcanoes that encircle the Central Plateau—Poás, Irazú, and Barva—and Volcano Turrialba southeast of Barva. Poás is active, and Irazú last erupted in 1963. Turrialba was becoming restless in 2008.

The fourth highland region is composed of the great chain of mountains from Cartago to the Panama border. They are the Talamanca Mountains and Cerro de la Muerte, which are of tectonic origin rather than volcanic. Included is Cerro Chirripó, the highest point in Costa

Crater of Poás volcano.

Talamanca Mountains, Cerro de la Muerte.

Rica at 15,526 feet. These mountains were formed when the Cocos Tectonic Plate pushed up from beneath the ocean onto the Caribbean Tectonic Plate about three to four million years ago. Much of this mountain range is protected as Tapantí NP (11,650 acres), Chirripó NP (123,921 acres), and La Amistad Costa Rica–Panama International Park (479,199 acres).

SPECIES DIVERSITY

Species diversity decreases with increasing elevation. For example, out of Costa Rica's 887 species of birds, about 130 species can be expected above 6,000 feet. About 105 species can be expected above 7,000 feet, about 85 can be found above 8,000 feet, and about 70 bird species can be expected above 9,000 feet. Comparable trends of lower diversity at higher elevations could also be expected among Costa Rica's invertebrate populations.

HUMBOLDT'S LAW

The South American explorer Alexander von Humboldt recognized an interesting relationship in tropical countries with high mountains. As one travels up a mountain, the average annual temperature decreases by 1 degree Fahrenheit for each increase of 300 feet in elevation. As one travels northward from the equator, the mean annual temperature decreases by 1 degree Fahrenheit for each sixty-seven miles of change in latitude. So an increase of 300 feet elevation on a mountain in the tropics is broadly comparable to traveling sixty-seven miles north. This relationship is portrayed in Figure 7 and is referred to as Humboldt's Law.

Some interesting changes in plant and animal life become apparent in travel up a mountain in the tropics that biologically resemble northward travel in latitude. The relationship of latitude and elevation becomes apparent at higher elevations because there are many temperate-origin plants and birds in the highlands. For example, the avifauna present at 8,000 feet on Costa Rica's mountains includes a higher proportion of temperate-origin thrushes, finches, juncos, and sparrows than is found in the tropical lowlands. It is interesting that the Painted Lady Butterfly (*Vanessa virginiensis*) is found in temperate regions of North America and also at the highest elevations in the paramo of Costa Rica.

Many plants of higher elevations in Costa Rica are in the same genera as

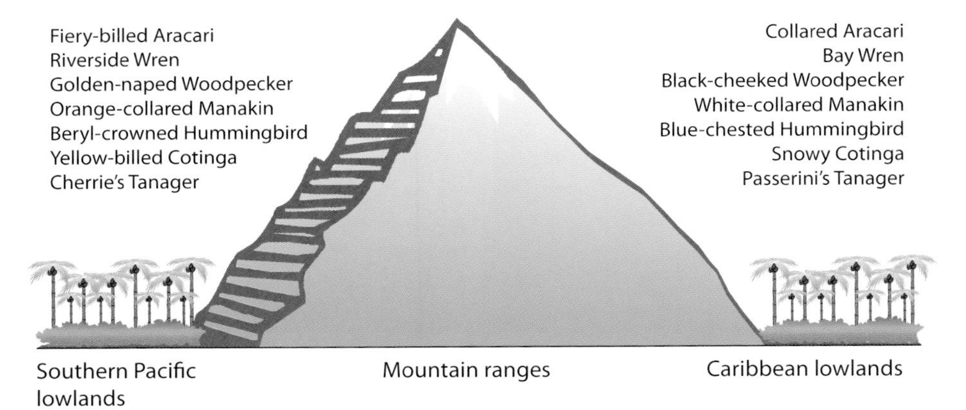

Figure 7. Closely related pairs of allopatric species of the Caribbean and Pacific lowlands that share a common ancestor but are now separated by Costa Rica's mountains.

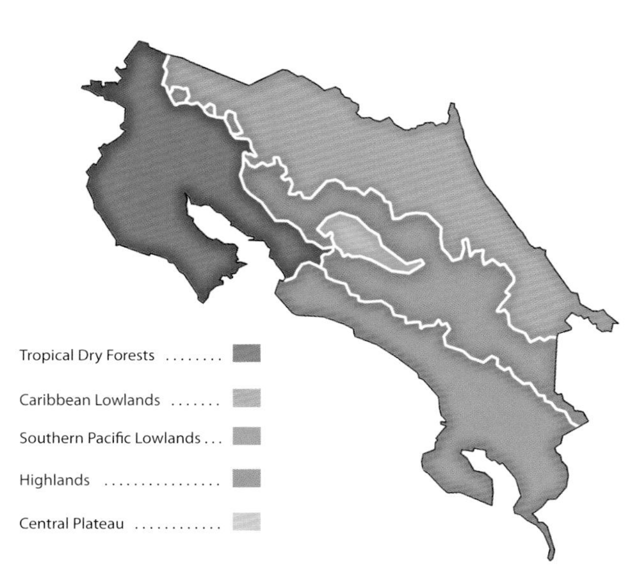

Figure 8. Five biological zones of Costa Rica. In addition, the country's entire coastline, beaches, and mangrove lagoons make up a sixth zone of biological importance.

plants found in the northern United States and Canada, including alders (*Alnus*), oaks (*Quercus*), blueberries (*Vaccinium*), blackberries (*Rubus*), bayberries (*Myrica*), dogwoods (*Cornus*), bearberries (*Arctostaphylos*), the Indian paintbrush (*Castilleja*), and the boneset (*Eupatorium*). Of course, in temperate areas there is a great deal more variation above and below the annual average temperature than in tropical areas, where there is little variation throughout the year; and in Costa Rica, there has never been a snowfall.

ELEVATIONAL ZONES

To understand the role that elevation plays in plant and animal distribution, it is useful to understand the main categories by which biologists classify elevations and how those zones relate to the highlands. These elevational zones are shown in Figure 8 and described below.

Tropical lowlands: The tropical lowland zone ranges from sea level to about 2,300 feet on the Pacific slope and 2,000 feet on the Caribbean slope. The lowland zone includes dry forests like those in Guanacaste as well as moist and wet forests of the southern Pacific and Caribbean lowlands.

PREMONTANE ZONE: This zone is called the "foothills" or "subtropical" zone and is also referred to as the "middle-elevation" zone. Some birds and other animals are found only in the foothills; examples are the Speckled and Silver-throated Tanagers. Two species of frogs are endemic to the microhabitats of Costa Rica's foothills, *Hylomantis lemur* in the Caribbean foothills and *Duellmanohyla rufioculis* in the Caribbean and Pacific foothills. This zone ranges from about 2,300 feet to 4,900 feet on the Pacific slope and 2,000 feet to 4,600 feet on the Caribbean slope. It could also be called the coffee zone because it is the zone in which the conditions are ideal for coffee production—and for human settlement.

LOWER MONTANE ZONE: The lower montane zone is part of the highlands. It includes the region from 4,900 feet to 8,500 feet on the Pacific slope and 4,600 feet to 8,200 feet on the Caribbean slope. One special habitat that occurs within this zone, and in upper levels of the premontane zone, is cloud forest. The cloud forest occurs roughly from 4,500 to 5,500 feet. A cloud forest, like that at Monteverde, is characterized by fog, mist, and high humidity as well as high precipitation—about 120 to 160 inches per year. The emerging problem of climate change is now causing the Monteverde forests to become drier and is placing cloud forest species of plants and wildlife in serious jeopardy.

Orchids, bromeliads, philodendrons, and dozens of other epiphytes grow in lush profusion among the branches of cloud forest trees. The Resplendent Quetzal is a well-known bird of the cloud forest, as well as lower montane and montane forests.

MONTANE ZONE: The montane zone ranges from 8,500 feet to 10,800 feet on the Pacific slope and from 8,200 feet to 10,500 feet on the Caribbean slope. Among plants of lower montane and montane zones are many of northern temperate origins: oak (*Quercus*), blueberry (*Vaccinium*), bearberry (*Arctostaphylos*), bamboo (*Chusquea*), alder (*Alnus*), bayberry (*Myrica*), magnolia (*Magnolia*), butterfly bush (*Buddleja*), elm (*Ulmus*), mistletoe (*Gaiadendron*), boneset (*Eupatorium*), dogwood (*Cornus*), Indian paintbrush (*Castilleja*), and members of the blueberry family (Ericaceae) like *Satyria, Cavendishia,* and *Psammisia*. Other conspicuous plants are *Oreopanax, Senecio,*

Montane wet forest, Cerro de la Muerte.

Cloud forest vegetation, Monteverde.

Miconia, Clusia, Bomarea, Giant Thistle (*Cirsium*), *Monochaetum,* Wild Avocado (*Persea*), Poor-Man's Umbrella (*Gunnera*), and tree ferns.

SUBALPINE PARAMO: Above the montane zone is the area above the treeline called the paramo. It has short, stunted, shrubby vegetation, including bamboo (*Chusquea*), many composites (like *Senecio*), and plants of South American origin from the Andes: a terrestrial bromeliad called *Puya dasylirioides* and a yellow-flowered composite with fuzzy white leaves called Lamb's Ears (*Espeletia*).

Coastal Beaches, Islands, and Mangrove Lagoons

The sixth biological zone, not portrayed on the map of biological regions (Fig. 8), includes all Pacific and Caribbean coastlines that extend from Nicaragua to Panama. This shoreline habitat consists of the beaches to the high-tide line and adjacent forests, offshore islands, rocky tidepools exposed at low tide, coral reefs, and mangrove lagoons. Among the more notable islands are Caño Island and Cocos Island. The only coral reef is at Cahuita NP.

Coastal beaches and river estuaries are extremely important as habitat for migratory shorebirds, seabirds, sea turtles,

Subalpine rainforest paramo.

Río Tárcoles estuary.

and mangrove-dependent wildlife species. Many of the birds, such as the Willet, Whimbrel, Wandering Tattler, and Ruddy Turnstone, winter along Costa Rica's beaches. Peregrine Falcons also winter along the coasts and prey on the shorebirds. Costa Rica's beaches are extremely important as nesting sites for Green Turtles on the Caribbean coast and for Ridley, Hawksbill, and Leatherback turtles on the Pacific coast. Among the mammals frequenting the beaches, mangrove lagoons, and estuaries are raccoons, coatis, tapirs, White-throated Capuchins, Silky Anteaters, river otters, and manatees (on the Caribbean coast).

Other important coastal habitats that are critically endangered by foreign beachfront developers and pollution are those of the mangrove lagoons and mangrove forests. These are important nurseries for fish and wildlife. Significant mangrove lagoons exist at Tamarindo, Playas del Coco, Gulf of Nicoya, Parrita, Golfito, Chomes, Boca Barranca, Quepos, and the 54,362-acre Mangrove Forest Reserve of the Ríos Térraba and Sierpe. They provide exceptional wildlife-watching opportunities during guided boat tours.

WILDLIFE OVERVIEW AND SPECIES COVERAGE

The fauna of Costa Rica includes thousands of birds, mammals, reptiles, amphibians, butterflies, moths, and other invertebrates. This diversity can be both overwhelming and inspiring to a nature enthusiast. Even the casual tourist is drawn to the tropical beauty and appeal of monkeys, motmots, and morphos.

The three volumes of the *Field Guide to the Wildlife of Costa Rica* series have been written to help visitors to Costa Rica learn more than just the names of species. They

Mangrove lagoon at Quepos.

are written to explain fascinating aspects of the ecology of common and unusual wildlife and the remarkable adaptations relating to predation, camouflage, and environment that are characteristic of many tropical organisms. These are the gee-whiz facts that make every day in a tropical forest a fun and scientific adventure, providing lifelong memories.

Series Coverage

During twenty-three trips to Costa Rica, Henderson birding tours have visited seventy-six sites. I have subsequently compiled an Excel spreadsheet listing more than 28,070 wildlife sightings from those tours. A total of 292 wildlife species were selected for the first edition of this book, in 2002. The expanded coverage of this three-volume set now includes more than 500 species, representing over 75 percent of the species sighted on our wildlife tours.

The first volume covers the country's diverse birdlife, including more than three hundred species accounts; it will leave you overwhelmed by the beauty, diversity, and fascinating ecological adaptations of birds in Costa Rica.

The second book in the series presents introductory information, species accounts, and about 160 colorful photos of more than 100 selected species of butterflies, moths, and other invertebrates that may be encountered during a visit to Costa Rica.

This third volume includes 115 selected species and 219 photos of some of the most conspicuous, colorful, and memorable mammals, amphibians, and reptiles in Costa Rica. This guide is different from most field guides, which try to cover all the species in a particular group of wildlife, like mammals or reptiles. Many of the species in those guides are so rare that casual tourists will never see them. This book instead provides a sampler of species, including many of the most conspicuous and interesting mammals, reptiles, and amphibians that are most appealing to tourists, based on my experience in leading tours in Costa Rica since 1987. Some of these are colorful creatures like leaf frogs, memorable mammals like monkeys and sloths, impressive reptiles like crocodiles, and amphibians with intriguing adaptations like poison dart frogs. Some accounts describe seldom-seen creatures like big cats, venomous snakes, and glass frogs, which are of keen interest to tourists who are curious about these well-known inhabitants of the rainforest.

The life-history information available for Costa Rica's wildlife varies greatly among species. Some have been well studied; others are largely unknown. The large amount of scientific literature that has been reviewed represents the best information available. Much information was obtained from Dr. Dan Janzen's monumental work *Costa Rican Natural History* (1983), as well as *The Amphibians and Reptiles of Costa Rica,* by Jay Savage (2002); *A Guide to Amphibians and Reptiles of Costa Rica,* by Twan Leenders (2001); *Leaf Frogs of Costa Rica,* by Brian Kubicki (2004); *Common Amphibians of Costa Rica,* by David Norman (1998); *Amphibians and Reptiles of La Selva, Costa Rica, and the Caribbean Slope,* by Craig Guyer and Maureen A. Donnelly (2005); *The Mammals of Costa Rica,* by Mark Wainwright (2007); *Mamíferos de Costa Rica,* by Eduardo Carrillo, Grace Wong, and Joel C. Sáenz (1999); *Murciélagos de Costa Rica,* by Richard K. LaVal and Bernal Rodríguez-Herrara (2002); *Mammals of the Neotropics: The Northern Neotropics,* by John F. Eisenberg (1989); *Neotropical Rainforest Mammals,* by Louise Emmons (1997); *A Field Guide to the Mammals of Central America and Southeast Mexico,* by Fiona A. Reid (1997); and *A Guide to the Carnivores of Central America,* by Carlos L. de la Rosa and Claudia C. Nocke (2000). All literature used is included in the bibliography.

Species Accounts

Each species account is preceded by the common name, scientific name, Costa Rican name, number of sightings, body length and weight (when available), geographic range, and elevational range. The number of sightings includes a summary of how many trips out of twenty-three Henderson tours the species has been encountered on, and the cumulative number of times that the species was recorded during those twenty-three trips. For example, "19/23 trips; 111 sightings" indicates that the species was encountered on nineteen out of twenty-three trips and recorded 111 times.

The accounts are also accompanied by distribution maps indicating the sites where each species was found on our various trips. Our Costa Rican tours have taken us to seventy-six sites, which are mapped and listed in Appendix B, along with details for each site and contact information for nearby lodges. Both elevations and geographic position system (GPS) readings of latitude and longitude were taken so these observations could be used to compile the distribution maps for the mammals. The pattern of dots within the biological zones portrays the general distribution of a species, but there are obviously many areas we have not visited where these species also occur. There is a seasonal bias to these sightings, as most were recorded in January and February. For amphibians and reptiles, distribution maps were compiled from the sightings on the Henderson birding tours and from distributional information cited in the Bibliography, including data from Savage (2002).

Photography

Photos by the author illustrate most of the species accounts in this volume, representing the best available from a personal collection of more than 60,000 Costa Rican and Latin American nature and wildlife images. Other photographers have contributed twenty-seven additional photos, including two images of White Tent Bats by Joanna Eckles, an Olingo photo by Robert Djupstrom, a Red Brocket Deer photo by Daniel H. Janzen, a Greater Fishing Bat photo by J. Scott Altenbach, twenty-two frog and toad images by Brian Kubicki, and Olive Ridley Turtle images by Pablo Vásquez Badilla. The postures, behavior, and natural colors displayed by these photos provide the best reference for nature enthusiasts. Paintings usually fail to convey the correct colors, proportions, and postures of the creatures portrayed because they are often painted from dead or faded museum specimens.

The author's photos were taken with a Canon 20D digital camera and Pentax 35 mm cameras (K-1000, SF-10, SF-1N, and PZ-1). Lenses included a Tamron 500 mm telephoto lens, Pentax 100–300 mm telephoto lens, and Sigma 400 mm APO telephoto lens (sometimes with a 1.4X Sigma teleconverter). Flash units included a Canon 580EX Speedlite, Pentax AF400FTZ for telephoto flash (used with a Lepp Project-a-Flash), and a Pentax AF240FT for macro photography. Fuji 100 Sensia I and Sensia II film was used for the 35 mm slides, and digital photos have been used since 2005.

Over 80 percent of the wildlife images included in this book were photographed in the wild, primarily in Costa Rica. Some have been photographed in the wild in other countries of Latin America, but they are the same species or subspecies that occur in Costa Rica. The remaining images were taken in captive settings either in Latin America or the United States. Some photos have been enhanced through the use of Adobe Photoshop to highlight identification marks and remove distracting background features.

SPECIES ACCOUNTS

Young Three-toed Sloth

MAMMALS

The fauna of Costa Rica includes at least 236 mammals. Considering the small size of the country, this represents a greater diversity of mammals than occurs in more northern temperate climates. For example, Minnesota is four times larger than Costa Rica and has just 80 mammal species.

A review of the mammals provides some revealing insights, such as that half are bats! In northern environments, few bats are present and all are insect eaters, but in tropical regions, bats fill many habitat niches and have diverse food habits. They eat insects, nectar, fruits, fish, and blood. Many bats are essential to the survival of tropical plants because they pollinate flowers or disperse the seeds. Additionally, because the complex tropical forest canopy provides numerous habitat niches, many mammals are adapted to living and traveling through the treetops to seek food, dens, and safety. These species include bats, opossums, monkeys, squirrels, sloths, small wild cats, and members of the weasel family like tayras.

Costa Rican mammals represent an unusual mix of species that have either temperate North American or tropical South American origins. These factors are explained in the Introduction under "Biogeography."

From a tourism and viewing standpoint, mammals can be difficult to see. Most are either crepuscular (active at dawn or dusk) or nocturnal. Many live in the dense forest and are difficult to spot. The most conspicuous mammals are monkeys, sloths, squirrels, and an occasional White-tailed Deer or agouti. The best opportunities for seeing mammals are while hiking early or late in the day or on escorted day or night drives, hikes, or boat tours with naturalist guides. When hoping to see the more elusive wild mammals, it is important to walk silently, stand or sit quietly for extended periods, and refrain from talking. In protected reserves like Corcovado NP, it is possible to see rare mammals like tapirs and peccaries because they tend to be more active during the day. This is also true for mammals in other national parks and at La Selva Biological Field Station, where agoutis and herds of Collared Peccaries roam during the day.

Costa Rican mammals make an incredible array of sounds that can bring life, suspense, and excitement to a tropical forest. An excellent CD, *Sounds of Neotropical Rainforest Mammals: An Audio Field Guide,* by Louise H. Emmons, Bret M. Whitney, and David L. Ross, Jr., is available from the Library of Natural Sounds at Cornell Laboratory of Ornithology. The roar of a jaguar, howling of howler monkeys, tooth clacking of peccaries, and birdlike whistle of a tapir are among the fascinating sounds included. The sounds of many mammals described in this book are recorded on that CD.

Forty-eight mammals have been selected for the species accounts in this chapter. The range maps accompanying the species

accounts include the specific locations where the mammals have been sighted on twenty-three trips by Henderson birding group members. Many of these species are very localized in the areas where they may be seen, so these maps help highlight those locations. The general range of the species can be implied from the pattern of the dots shown. The mammals included here represent some of the most conspicuous and abundant mammals as well as some of the more rare and seldom-seen species that add a sense of wonder, mystery, and excitement to Costa Rica's tropical forests. Even if they are unseen, they are probably there, watching you.

Mantled Howler Monkey peeking from the canopy

COMMON OPOSSUM

The Common Opossum is the largest of nine opossums in Costa Rica. Primarily nocturnal, this adaptable marsupial occurs in pristine rainforests, coffee plantations, farms, cities, and gardens. Its range includes Caribbean lowlands, dry forests of Guanacaste, and southern Pacific lowland forests near the Panama border.

The size of a large house cat, the Common Opossum has grizzled gray or black fur. The cheeks are a creamy to yellowish color, in contrast to the white cheeks of the Virginia Opossum (*D. virginiana*), which also occurs in Guanacaste. Foods include natural and cultivated fruits, seeds, insects, crayfish, snails, mice, snakes, lizards, eggs, small birds, and even fish.

Females bear about six young twice per year, primarily in February and July. The young remain in the female's pouch for about sixty days after being born. When approached, this opossum growls, hisses, and bites an attacker; it does not play 'possum like the Virginia Opossum does. Predators include coyotes, feral dogs, boa constrictors, tayras, ocelots, pumas, and jaguars. It is reported to be immune to the bites of venomous snakes.

This opossum may be seen with the aid of spotlights on night excursions in farm or ranch groves in Guanacaste, at La Selva Biological Field Station, in Manuel Antonio NP, and in the San Isidro del General area.

Didelphis marsupialis

Costa Rican names: *Zarigüella; raposa; zorro pelón; zorra mochila.*

8/23 trips; 14 sightings.

Total length: 25.4–40.0 inches, including 10.0–21.1-inch tail.

Weight: 1 pound 7 ounces–4 pounds 10 ounces (665–2,090 grams).

Range: Southern Texas to northern Argentina.

Elevational range: Sea level to 4,900 feet, with some records up to 7,300 feet.

Common Opossum adult

Central American Woolly Opossum adult

CENTRAL AMERICAN WOOLLY OPOSSUM

The Central American Woolly Opossum is a distinctive marsupial of mature moist and wet forests, older second-growth forests, and forest edges. Its back is orange, its ears are large and pink, and its long tail is covered with fur for about one-third to one-half its length. There are no spots above the eyes, as on the four-eyed opossums. The Costa Rican name, *Zorro de balsa,* refers to this nocturnal animal's habit of visiting balsa trees at night to drink nectar from the flowers. The species also eats figs and the fruit of *Piper* species. When spotted at night with the aid of lights, the eyes are bright red.

In contrast to other opossums, which are known for slow and lumbering movements, the Woolly Opossum is slender, agile, and quick. It can leap from one branch to another, run along slender telephone wires or tree branches, and hang by its tail while catching and eating passing moths. It is so light-footed and adept at climbing out onto slender tree branches that it can gather fruits, seeds, flowers, and small creatures that other opossums could not reach. This opossum does not play dead when threatened. It will leap at its attacker and try to bite.

This opossum becomes sexually mature at seven to nine months of age. It is easy to tell the sexes apart, because the males have a conspicuous blue scrotum. After a gestation period of about two weeks, two to four young are born. At first the female carries them in a pouch, like other opossums, but the care of these young is extended. After about two and a half months in the pouch, the young are placed in a nest, where the mother continues to nurse them until the age of four months.

The Central American Woolly Opossum is nocturnal. Spotlights must be used to look for them at night along roadsides or trails where the presence of *Cecropia* trees might give the viewer an opportunity to see them. They are found in the dry, moist, and wet lowlands of both the Caribbean and Pacific slopes. They can be seen in Santa Rosa and Tortuguero NPs and on the grounds of La Cusinga Rainforest Lodge.

Caluromys derbianus

Costa Rican name: *Zorro de balsa.*

2/23 trips; 2 sightings.

Total length: 23.1–29.9 inches, including 15.5–19.1-inch tail.

Weight: 10.4 ounces (295 grams).

Range: Veracruz, Mexico, to western Ecuador.

Elevational range: Sea level to 8,000 feet.

GRAY FOUR-EYED OPOSSUM

Philander opossum

Costa Rican name: *Zorro de cuatro ojos.*

3/23 trips; 3 sightings.

Total length: 21.0–24.0 inches, including 11.1–11.8-inch tail.

Weight: 9.8 ounces–3 pounds 1 ounce (263–1,400 grams).

Range: Tamaulipas, Mexico, to northeastern Argentina.

Elevational range: Sea level to 4,800 feet.

The Gray Four-eyed Opossum gets its name from two white spots above its eyes, giving it a four-eyed appearance. The size of a small house cat, this marsupial has short grayish fur that is not as grizzled as the fur of Common or Virginia Opossums. The very long tail is dark and furred at the base and light and bare toward the tip. The similar Brown Four-eyed Opossum (*Metachirus nudicaudatus*) can be distinguished by a buff (not white) spot over each eye, brownish color over the back, and a longer tail than that of the Gray Four-eyed Opossum. The diet consists of wild fruits, nectar, fish, crabs, frogs, mice, and insects. It is also encountered on farms, where it eats fruits and grains. Predators include jaguars, pumas, ocelots, tayras, and boa constrictors.

This marsupial produces a litter of four or five young twice each year. Kits stay attached to the teats of the female for at least sixty days. The young reach sexual maturity at seven months of age.

This opossum inhabits dry and moist lowland forests of both the Caribbean and Pacific slopes near streams and swampy areas. It may be encountered at night in trees or on the ground at La Pacífica, La Selva Biological Field Station, Tortuguero NP, or Selva Verde in the Caribbean lowlands.

Gray Four-eyed Opossum adult

SAC-WING BAT (GREATER WHITE-LINED BAT)

Saccopteryx bilineata

Costa Rican names: *Murciélago de saco; murciélago de listas.*

7/23 trips; 8 sightings.

Total length: 2.1–2.2 inches, including 0.8-inch tail.

Weight: 0.22–0.33 ounce (6.2–9.3 grams).

Range: Colima, Mexico, to Brazil and Trinidad and Tobago.

Elevational range: Sea level to 1,650 feet.

The Sac-wing Bat is a common insect-eating bat in Pacific and Caribbean lowland forests, including dry forests of Guanacaste. Often more conspicuous than other bats, it roosts during the day on the sheltered buttresses of large trees such as strangler figs (*Ficus*) and kapok (*Ceiba*) trees. Roosts may also be encountered under large tree limbs and under the eaves of porch roofs. As they roost upside down, their heads are uplifted at a sharp angle that is characteristic of this genus. This insectivorous bat is both crepuscular and nocturnal. It usually forages near streams, moist habitats, and forest clearings.

This bat has short dark brown to black fur and two unbroken wavy white lines along the back. It occurs in harems of one to nine females attended by a larger male. The name "Sac-wing" comes from scent glands under the wings that the male uses to attract females and to intimidate male bats. A male will hover in front of his harem while chirping to them and wafting his bat scent toward them. A female produces one young per year. Young can fly when only two weeks of age, but they will continue nursing for several months.

This bat may be observed under thatched eaves at Selva Verde Lodge, in abandoned settlers' cabins at La Selva Biological Field Station, in La Pacífica, in historic buildings at Santa Rosa NP, and at the Sirena Biological Station in Corcovado NP.

Sac-wing Bat

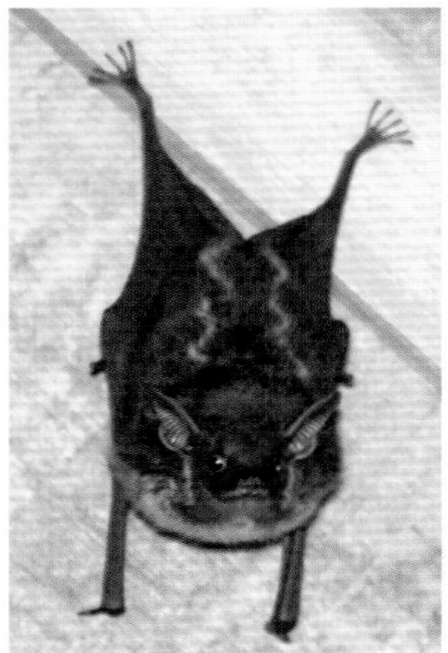

Sac-wing Bat, dorsal view showing unbroken wavy lines on the back

BRAZILIAN LONG-NOSED BAT

The Brazilian Long-nosed Bat looks much like the Sac-wing Bat, but the two wavy white lines on its back are broken. It is typically seen hanging upside down in vertical rows on the underside of leaning palm tree trunks or tree limbs overhanging rivers and canals in tropical lowlands. Colonies include ten to twenty-four individuals, with several larger males among the smaller females. They do not form harems, as do Sac-wing Bats. Feeding is done over water.

Sightings can be made in the Caribbean lowlands on palm trunks along the water's edge on the Río Frío, in Caño Negro NWR, along the Río Sarapiquí, at La Laguna del Lagarto Lodge, at Tortuga Lodge, and in Tortuguero NP. In the Pacific lowlands they can be encountered on the grounds and in mangrove lagoons of La Ensenada Lodge, in the Damas Island mangrove lagoons near Quepos, in the Villa Lapas courtyard, and in Esquinas Lodge courtyard and grounds.

Rhynchonycteris naso

Costa Rican name: *Murciélago de trompa.*

6/23 trips; 11 sightings.

Total length: 1.5–1.7 inches.

Weight: 0.13–0.14 ounce (3.8–3.9 grams).

Range: Veracruz, Mexico, to southeastern Brazil.

Elevational range: Sea level to 1,000 feet.

A colony of twelve Brazilian Long-nosed Bats on a palm tree trunk

Brazilian Long-nosed Bats at a daytime roosting site

SMOKY BAT

Cyttarops alecto

Costa Rican name: *Murciélago.*

1/23 trips; 1 sighting.

Total length: 1.8–2.2 inches.

Weight: 0.21–0.24 ounce (6–7 grams).

Range: Eastern Nicaragua and eastern Costa Rica. Disjunct populations in Guiana and northeastern Brazil.

Elevational range: Sea level to 1,000 feet.

The rare Smoky Bat (also referred to as the Short-eared Bat), is all gray and is known only from the tropical lowlands of northeastern Costa Rica. Its distribution poses an interesting puzzle. A northern population exists in southeastern Nicaragua and northeastern Costa Rica, and disjunct populations are also known in the Guianas and northeastern Brazil. In the region between those areas, however, no Smoky Bats are known.

An insect-eating species, this bat roosts during the day in the central arching portions of palm fronds where the palms have a relatively open understory, for instance, in a courtyard or plantation. The author encountered six of these bats in the courtyard of Tortuga Lodge in January 1997.

After sunset these bats fly out to feed. They are so rare that most details of their life history and reproduction are unknown.

Rare Smoky Bats roosting under a palm frond

Smoky Bats under a palm frond at Tortuga Lodge

TENT-MAKING BAT

Tent-making Bats get their name from the way they bite the main veins of long leaves, like *Heliconia* or palm leaves, so that the leaves droop to create a tentlike waterproof shelter. This protects them from rain in areas that receive up to 200 inches of precipitation per year. These bats sleep in their leaf shelters in tightly packed colonies ranging from two to fifty-nine individuals. The species is characterized by two prominent white stripes on the face and a leaflike nose. Females roost in separate maternity colonies, where they bear their young. Each female gives birth to two young. Pups are believed to be born from January through April. They stay with the female for one month.

The Tent-making Bat is found in moist and wet Caribbean and Pacific lowlands. Although primarily a fruit-eating species, it also eats insects. Foods include fruits of figs and piper plants. This bat may be encountered at Tortuga Lodge, at Villa Lapas, and in the vicinity of La Selva.

Uroderma bilobatum
Costa Rican name: *Murciélago.*
5/23 trips; 7 sightings.
Total length: 2.4 inches.
Weight: 0.46–0.70 ounce (13–20 grams).
Range: Southern Mexico to southeastern Brazil.
Elevational range: Sea level to 4,300 feet.

Tent-making Bat colony

A maternity colony of Tent-making Bats with light gray young clinging to the bellies of their mothers

Heliconia leaf in use by Tent-making Bats

GREAT FRUIT-EATING BAT

Artibeus lituratus

Costa Rican name: *Murciélago frutero.*

1/23 trips; 1 sighting.

Total length: 3.4–4.0 inches.

Weight: 1.8–3.0 ounces (50–86 grams).

Range: Sinaloa, Mexico, to northern Argentina.

Elevational range: Sea level to 5,100 feet.

As the name implies, this is a large bat that primarily eats fruits of rainforest trees and shrubs. It fills an important ecological role as a disperser of the seeds from rainforest plants. The diet includes figs, the fruits of *Piper* species, flower nectar, pollen, and insects that are taken on the wing. Fruit is plucked and taken in the bat's mouth back to the roost, where it is consumed.

This bat is one of several leaf-nosed species with a pair of white parallel stripes from the nose along the top of the head. It is dark brown on the back, and the tail membrane is covered with fur. Similar species include the grayish-brown Jamaican Fruit-eating Bat (*Artibeus jaimaicensis*), which has no tail, and the San Jose Fruit-eating Bat (*A. intermedius*), which is uncommon and found primarily in dry forests of Guanacaste.

Adaptable and widespread, the Great Fruit-eating Bat occurs in habitats ranging from dry forests to wet forests. Day-roosting sites may include caves, hollow trees, and crevices in rock formations, but it also can be observed sheltered under palm leaves. The bats shown here were encountered in the courtyard of Villa Lapas on the Pacific coast.

Great Fruit-eating Bats

WHITE TENT BAT

Ectophylla alba
Costa Rican name: *Murciélago blanco.*
1/23 trips; 1 sighting.
Total length: 1.4–1.7 inches.
Weight: 0.3 ounce (7.5 grams).
Range: Honduras to northwestern Panama.
Elevational range: Sea level to 2,100 feet.

The White Tent Bat is one of the most distinctive bats in Costa Rica—it is all white. The only other white bat in the country, *Diclidurus albus,* does not have a leaf nose and does not make tents for shelter. The White Tent Bat is leaf-nosed and lives in colonies of about four to eight individuals that build leaf tents. They nip the veins of horizontally suspended leaves of banana (*Musa*), *Heliconia*, or *Calathea* plants, causing the leaf to collapse into a tentlike structure that provides protection from rainfall. These leaves are often within a few feet of the ground. Preferred habitats include primary and secondary rainforests and plantations.

The White Tent Bat is an uncommon, fruit-eating species. It flies out at night in search of fruits like figs (*Ficus*). The fruits are carried back to the roost, where they are consumed. Sometimes it is possible to spot waste fruits that have been dropped by roosting bats under a tent structure.

Single young are born in April. When females begin to use the tent as a nursery, the males move to another tent. Females will nurse the young of other females in their nursery group.

These bats are found in a relatively small range from Honduras to Panama, including the Caribbean lowlands of Costa Rica. They occur at the Tirimbina Research Station (operated by the Milwaukee Public Museum) and are regularly observed on the island in the lagoon. They can be observed during guided walks at that site. Be especially careful not to disturb these bats from their roosts.

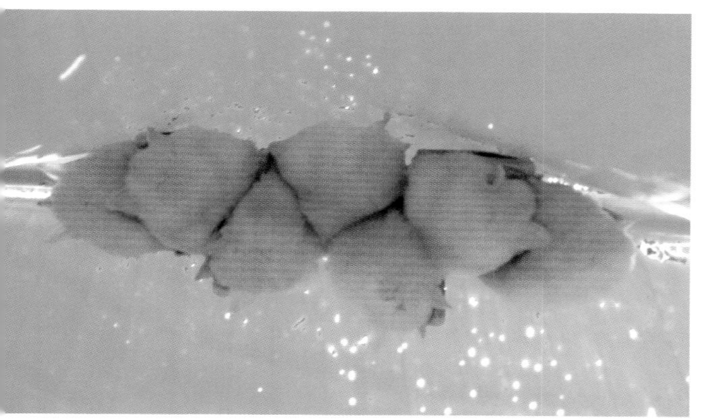

White Tent Bat colony. Photo © Joanna Eckles

White Tent Bats under a *Heliconia* leaf. Photo © Joanna Eckles

GREATER FISHING BAT (GREATER BULLDOG BAT)

Greater Fishing Bat catching minnow. Photo © J. Scott Altenbach

Noctilio leporinus

Costa Rican name: *Murciélago pescador.*

3/23 trips: 3 sightings.

Length: 3.0–4.2 inches.

Weight: 1.7–2.7 ounces (49–78 grams).

Range: Mexico to northern Argentina.

Elevational range: Tropical lowlands below 700 feet.

Among the more than 100 bat species in Costa Rica, the Greater Fishing Bat is one of the largest and most memorable. It is also called the Greater Bulldog Bat, because its face resembles a bulldog. It is also called a fishing bat because of its peculiar fishing habits. After dark it flies over brackish, fresh-, or saltwater lagoons, using its sonar to locate minnows or insects at the water's surface. With a wingspread of two feet or more, this conspicuous bat is sometimes attracted to fish under dockside security lights at rainforest lodges, where topminnows or other prey species come to feed on insects attracted to the light. It may consume thirty to forty minnows per night.

Certain features of this bat are particularly suited to its fishing style. Its long legs trail above the water as it uses sonar to locate small fish. Unlike any other bat, its claws are curved forward, allowing it to grasp the fish. If a ripple is detected, it dips its feet into the water and drags its claws, snatching up its prey and transferring it to its mouth. Usually the bat will then fly to a perch in a tree, where it swallows the fish.

The Greater Fishing Bat lives in colonies of up to several hundred individuals; they inhabit hollow trees and caves along rivers and backwater canals in tropical lowlands. This bat has also been known to roost under bridges and in buildings. Females form maternity colonies, where each bears one young per year.

These bats can be observed along the Puerto Viejo River in the vicinity of La Selva and also in canals of Tortuguero NP. They were observed feeding at the dock of Tortuga Lodge in 1990 and 1991, where over a dozen bats regularly fed on topminnows attracted to the security lights at night. The activity of the bats, however, attracted a Black-and-White Owl, and over an extended period, the owl captured and ate all of the bats. It would perch in a palm tree above the dock, watching the circling bats and learning to anticipate their fishing maneuvers. The author watched as the owl captured the last bat of this foraging group in January 1991.

VAMPIRE BAT

Desmodus rotundus

Costa Rican names: *Vampiro, mordedor.*

1/23 trips; 1 sighting.

Total length: 3.0–3.5 inches.

Weight: 0.70–1.52 ounces (19–43 grams).

Range: Tamaulipas, Mexico, to northern Argentina.

Elevational range: Sea level to 2,700 feet.

The Vampire Bat has a sinister reputation because of its feeding preference. Most bats eat insects, nectar, or even fish, but the Vampire Bat eats blood. Don't let the presence of Vampire Bats keep you from visiting Costa Rica, though; the author knows of no instances of tourists being bitten by these bats.

This bat flies out about two hours after midnight in search of sleeping cattle, horses, and even chickens. It may land on the victim or on the ground and walk, hop, or run to the victim. Large mammals are often bitten on the neck. The saliva contains an anticoagulant that allows blood to flow from the bite wound. The bat laps up the blood, about 15 milliliters per day. Sometimes another Vampire Bat will take over at a wound site when the first bat is finished. Vampire Bats prefer to feed on dark nights rather than in moonlight. They may be abundant around cattle ranches where they have a significant food supply. In the daytime, they roost in hollow trees, in caves, and even under bridges. A colony may contain up to a couple thousand individuals.

Reproduction may occur at any time of year. A single young is born after about seven months of gestation. Females nurse their young for seven to ten months. After three months, the females will fly out to feed and return to regurgitate blood to the young.

In the past, hatred of Vampire Bats caused many harmless and beneficial bats to be killed. Vampire Bats can cause a loss of vigor and health among livestock. Although they can carry rabies, rabies is rare in Costa Rica. Now problem colonies of Vampire Bats can be controlled by treating livestock with an anticoagulant that is ingested by the bats. When they fly back to the colony and participate in mutual grooming, the anticoagulant is ingested by the bats, and the colony numbers are significantly reduced without affecting other bats.

The only evidence of Vampire Bats encountered on twenty-three Henderson tours in Costa Rica was a horse that was observed with blood on its neck one morning in Guanacaste near the Nicaraguan border.

Vampire Bat adult

Mantled Howler Monkey, howling

MANTLED HOWLER MONKEY

The lionlike roars of the Mantled Howler Monkey make it the most conspicuous of four wild primates in Costa Rica's forests. The sound can be heard up to a mile away. Adults have prehensile tails that aid in traveling through the treetops. They also provide a secure grip as the monkey reaches for the leaves, fruits, and flowers that make up its diet. The black fur is highlighted by a fringe of long copper-colored hair on the sides and lower back, giving this howler the name "mantled."

Howlers live in troops of two to over forty individuals, averaging about thirteen. At La Pacífica in Guanacaste, densities approach one troop per seventy-five acres. Young monkeys reach sexual maturity at three to three-and-a-half years of age. The youngest sexually mature male and female in a troop are the socially dominant leaders. They maintain their dominance for thirteen to forty-eight months before being displaced by younger individuals. Only about 30 percent of young howler monkeys survive their first year, but some individuals may live up to twenty-five years. The gestation period is about 180 days and young stay with the mother for twelve to fifteen months.

This monkey lives in mangrove, dry, and wet forests

Mantled Howler Monkey, adult female and young sucking its thumb

from sea level up to montane cloud forests. Tolerant of humans, it occupies second-growth forests and even ranch woodlots. Many tropical forest wildlife species feed on nutritious fruits or seeds, but the Mantled Howler Monkey primarily eats roughage—young leaves of a wide variety of tropical trees and vines that are low in nutrition but available in great abundance. A few fruits and flowers are also eaten. Many tropical plants contain toxins like alkaloids, cyanide, or strychnine, so howler monkeys learn to select plants with low concentrations of such chemicals. Some trees increase the concentrations of these toxins in response to feeding by the monkeys, so the monkeys need to keep changing their diet to find plants low in toxins.

Howlers are readily seen at close range in the dry forests of Guanacaste. This includes Santa Rosa, Guanacaste, and Palo Verde NPs; Lomas Barbudal BR; La Ensenada Lodge; La Pacífica; Hacienda Solimar; and forests near Tamarindo, Brasilito, and Sugar Beach. They can also be seen from boats in the canals at Tortuguero NP and in moist and wet forests of the Pacific slope, including Carara NP, Sirena Biological Station, private lodges in the vicinity of Corcovado NP, and Tiskita Jungle Lodge. They are often heard but seldom seen in the cloud forests of Monteverde. While watching howler monkeys, keep an eye on the males. They may circle around you, get overhead, and urinate on you. This is a defense against predators.

Allouata palliata

Costa Rican names: *Congo*; *mono negro*.

20/23 trips; 141 sightings.

Total length: 41.1–50.4 inches, including a 21.5–25.8-inch tail.

Weight: 6 pounds 13 ounces–21 pounds 10 ounces (3.1–9.8 kilograms).

Range: Veracruz, Mexico, to western Colombia and Ecuador.

Elevational range: Sea level to 7,100 feet.

Mantled Howler Monkey eating a leaf

A troop of Mantled Howler Monkeys

CENTRAL AMERICAN SPIDER MONKEY (BLACK-HANDED SPIDER MONKEY)

Ateles geoffroyi
Costa Rican names: *Mono araña; mono colorado.*
16/23 trips; 46 sightings.
Total length: 37.0–57.9 inches, including 25.0–33.1-inch tail.
Weight: 13 pounds 3.5 ounces–19 pounds 13.2 ounces (6–9 kilograms).
Range: Veracruz, Mexico, to southeastern Panama.
Elevational range: Sea level to 7,200 feet.

Central American Spider Monkey

The Central American Spider Monkey has become endangered throughout its range because of habitat loss and killing of adults so the young can be captured for the illegal pet trade. Adults weigh about the same as a howler monkey, but the stocky build of howler monkeys makes them appear heavier. The Central American Spider Monkey has blond to reddish-brown fur over the back and sides. The hands and feet are black. Three subspecies occur in Costa Rica: *A. g. ornatus* in the Caribbean lowlands and central highlands, *A. g. frenatus* in Guanacaste, and *A. g. panamensis* in the lowlands and middle elevations of southwestern Costa Rica.

Spider monkeys, active throughout the day, move gracefully through treetops in search of ripe fruits. They also eat seeds, flowers, leaves, buds, and some small animals. Although a troop may consist of more than twenty individuals, the monkeys separate into several foraging groups during the day. At dusk, the monkeys gather to sleep. The home territory of spider monkeys ranges from 153 to 284 acres.

After maturing at the age of four to five, spider monkeys may live up to twenty-seven years. Females produce their first young at five to seven years of age, and then they give birth to one young at intervals of seventeen to forty-five months. Infant monkeys are cared for by the mother for over a year.

Spider monkeys are found throughout Costa Rica's lowlands, including dry forests of Guanacaste and moist and wet lowland forests of the Caribbean and southern Pacific slopes. Look for spider monkeys along the canals of Tortuguero NP, La Selva Biological Field Station, and Santa Rosa and Corcovado NPs, including the Sirena Biological Station.

Central American Spider Monkey, adult posing for the author

Central American Spider Monkey, adult female and young

Central American Spider Monkey

WHITE-THROATED CAPUCHIN (WHITE-FACED CAPUCHIN)

Cebus capucinus

Costa Rican names: *Mono cariblanco; mono cara blanca; mico.*

20/23 trips; 49 sightings.

Total length: 27–39.5 inches, including 13.8–21.7-inch tail.

Weight: 5 pounds 14.4 ounces–8 pounds 8.3 ounces (2.6–3.8 kilograms).

Range: Belize to northern and western Colombia.

Elevational range: Sea level to 8,200 feet.

Most people recognize this primate as the organ-grinder's monkey; like other Neotropical primates, however, this monkey belongs in the forest, not at the end of someone's leash. Smaller than howler and spider monkeys, the White-throated Capuchin has a prehensile tail and is characterized by a black body with a white to yellowish face, throat, and shoulder region. It is believed to be the most intelligent of all Costa Rican primates.

These monkeys are found in forests ranging from dry and mangrove forests in Guanacaste to moist and wet lowland forests of the Caribbean and southern Pacific coasts. They even occur up to the level of montane forests in the Talamanca Mountains. Capuchins are diurnal and feed on fruits, nuts, shoots, buds, flowers, caterpillars, cicadas, beetles, and ants. They are also carnivorous and will hunt for young squirrels and raid nests in search of nestling birds and young coatis. In contrast to spider monkeys, which feed primarily in treetops, this monkey forages from ground level to the upper canopy. White-throated Capuchins reach sexual maturity at three years of age and may live more than forty years. Females have a single young at approximately nineteen-month intervals. Infants are weaned at one year of age. This monkey lives in groups of five to thirty-six individuals that defend home ranges of 79–210 acres. They benefit forests by pollinating flowers and dispersing seeds in their droppings.

Capuchins are easily seen in most lowland forests of Costa Rica, including Santa Rosa, Guanacaste, Palo Verde, Tortuguero, Corcovado, Braulio Carrillo, Manuel Antonio, and Carara NPs. They can be seen in cloud forests of Monteverde, in the montane forests of Savegre Mountain Lodge, and even in mangrove forests near Quepos.

White-throated Capuchin, eating

White-throated Capuchin adult

Pair of White-throated Capuchins, threat display

RED-BACKED SQUIRREL MONKEY

Saimiri oerstedii

Costa Rican names: *Tití, mono tití.*

10/23 trips; 20 sightings.

Total length: 24.9 inches, including 14.3-inch tail.

Weight: 1 pound 5.1 ounces–2 pounds 1.5 ounces (600–950 grams).

Range: Endemic to southern Pacific coast of Costa Rica and adjacent Panama lowlands.

Elevational range: Sea level to 1,545 feet.

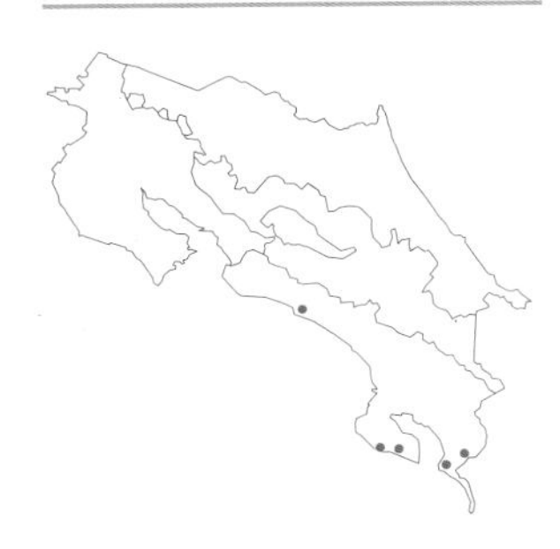

The beautiful Red-backed Squirrel Monkey is the smallest and rarest primate in Costa Rica. Its body is about the size of a large North American Fox Squirrel. It is the only monkey in this country without a prehensile tail. It also poses a great distributional puzzle. This monkey is separated from South American squirrel monkeys by the Andes Mountains in Colombia. It is suspected that pre-Columbian Indians from South America had trade routes along the Pacific Coast and brought monkeys as pets to native tribes in Central America. Some squirrel monkeys apparently escaped and adapted through natural selection over thousands of years until they became a separate species.

There are two subspecies of this monkey in Costa Rica that are distinctive and geographically separate. That of the Osa/Golfo Dulce/Tiskita region near Panama (*Saimiri oerstedii oerstedii*) has a black cap and is reddish over the back, shoulders, and flanks. The subspecies of Quepos and the Manuel Antonio NP region (*Saimiri oerstedii citrinellus*) has a gray cap and a reddish saddle over the back. It is grayish on the shoulders and flanks.

These monkeys move through treetops with long, graceful leaps as they search for fruit, seeds, leaves, birds' eggs, and insects. The long tail helps the monkey keep its balance during leaps. The habitat includes moist and wet Pacific lowland forests from Manuel Antonio NP to the Osa Peninsula, including Corcovado NP and lowlands east of Golfo Dulce, into western Panama.

Red-backed Squirrel Monkey, southern *oerstedii* race. Note black cap and red back.

Red-backed Squirrel Monkey, southern *oerstedii* race, with black cap

This diurnal monkey is easily viewed in groups of ten to sixty-five individuals in the treetops. The home range varies from forty-two to ninety-nine acres. Red-backed Squirrel Monkeys mate in January and February and have young after 170 days. Females have their first young at two years and produce one young about every thirteen to fourteen months thereafter. Squirrel monkeys can live up to twenty-one years. Predators include boa constrictors, tayras, Collared Forest-Falcons, and hawk-eagles.

This endangered primate is present in Corcovado NP, including the Sirena Biological Station. The best place to view this monkey is near the cabins at Tiskita Jungle Lodge, at Rancho Casa Grande, and at Manuel Antonio NP. As with howler monkeys, be careful when observing them at close range. They may get overhead and defecate on you! The author can verify that they have uncanny accuracy.

Red-backed Squirrel Monkey, northern *citrinellus* race, with gray cap

Red-backed Squirrel Monkey, northern *citrinellus* race. Note gray cap and pale rusty back.

Squirrel monkey in mid-leap at Tiskita Jungle Lodge

Red-backed Squirrel Monkey, showing long, nonprehensile tail

Giant Anteater, walking on its knuckles to accommodate its long claws

Myrmecophaga tridactyla
Costa Rican names: *Oso hormiguero gigante; Oso caballo.*
0/23 trips; 0 sightings.
Total length: 39.4–51.2 inches.
Weight: 48.4–85.8 pounds (22–39 kilograms).
Range: Guatemala and southern Belize to northern Argentina.
Elevational range: Sea level to about 1,000 feet.

GIANT ANTEATER

The Giant Anteater is one of the largest mammals in Costa Rica, and it is also the rarest. Few sightings have been recorded since 1984, when one was seen in Santa Rosa NP. This mammal may never have been abundant in the country, but it is so impressive that it was apparently shot on sight by early settlers as a novel trophy. The author recalls seeing the hide of a Giant Anteater tacked on the wall of an office at the international airport in San José in 1969, when the airport was still located at the current site of the Sabana Parque, but the origin of that hide was unknown.

The Giant Anteater is usually solitary and ranges over large areas of dry forest, grasslands, or tropical forests in search of ant or termite colonies, which it rips open with its huge claws. It then flicks out its two-foot tongue up to 150 times per minute to lap up the ants or termites. Ants of the genera *Componotus* and *Solenopsis* are preferred over termites, Leaf-Cutter Ants, and Army Ants. It may eat 30,000 ants in a single day.

The Spanish name for this species, *oso hormiguero*, means "anteater bear." It is not a bear, but when approached it can rear up on its hind legs like a bear and challenge its attacker with formidable claws. Because of the size and position of those claws, the Giant Anteater walks with its claws folded backward, actually walking on its knuckles.

Males require a large home range that may vary from 670 to 6,200 acres, which they defend against other males. Females have smaller home ranges. After a gestation period of six months, a single young is born. It is weaned after four to six weeks and thereafter will ride on the back of the mother. Giant anteaters can live from sixteen to twenty-six years.

Because of the apparent extirpated status of this impressive species, this would be a prime candidate for reintroduction in Costa Rica. This species is still present in good numbers in the llanos of Venezuela. Perhaps animals could be captured there and transplanted to suitable habitat in Costa Rica.

Giant Anteater

TAMANDUA (LESSER, BANDED, OR COLLARED ANTEATER)

Tamandua mexicana

Costa Rican names: *Tamandua; oso jaceta; hormiguero.*

6/23 trips; 8 sightings.

Total length: Up to 44.6 inches, including 21.4-inch tail.

Weight: 7 pounds 1 ounce–18 pounds 11 ounces (3.2–8.5 kilograms).

Range: Veracruz, Mexico, to Peru.

Elevational range: Sea level to 4,900 feet.

The size of a raccoon, the Tamandua is the most common of three anteaters in Costa Rica. When approached by a dog or other predator, this mammal stands on its hind legs, balancing with its long prehensile tail and holding its muscular forelegs in a pose like a boxer. If an animal approaches too closely, the Tamandua will slash downward with its long, sharp claws.

The head and body are tan to light brown, with black over the back and a black stripe extending forward over each shoulder. The forelegs have sharp, well-developed claws used to tear open nests of ants and termites. The slender snout and long, sticky tongue are used to probe for ants, termites, and occasional bees that constitute its diet.

Tamanduas live in dry, moist, and wet forests of the Caribbean and Pacific slopes. Some live in trees, and their diet comprises mostly ants. Others forage primarily for termites on the ground. Some are active in daytime; others forage at night. Tamanduas are solitary and occupy a territory of about 185 acres.

One young is born to an adult female each year and stays with the female until it is about half grown. When moving from one location to another, the young rides on the back of the mother. It dismounts to feed when the mother locates ants or termites.

The low frequency of Tamandua sightings is partially a reflection of how difficult it is to spot small mammals in the rainforest canopy. It is much easier to see them in the sparse foliage of the dry forests in Guanacaste. This anteater may be encountered at Palo Verde, Guanacaste, Tortuguero, Corcovado, and Santa Rosa NPs; in Caño Negro NWR; at La Selva Biological Field Station; and in the Osa Peninsula, including the Corcovado Lodge Tent Camp and along the nearby trail to the Río Madrigal.

Tamandua adult, feeding on ants in a bullhorn acacia tree

Tamandua, sleeping in a bullhorn acacia tree

Silky Anteater adult, defensive posture

SILKY ANTEATER

The Silky Anteater has the fascinating local name *serafín*, meaning "little angel." A beautiful little animal with golden silky fur, it is nocturnal and arboreal. In lowland rainforests it lives amid thick tangles of vines, where it searches at night for hollow stems that it slices open with sharp claws. It quickly laps up ants from the opening, and then it moves to another site. It may eat up to 3,000 ants every night.

During the day this small anteater curls up amid the vines to sleep. At a distance it looks like a shiny golden tennis ball. It conserves energy by going into a state of torpor in which its body temperature drops and its respiration rate decreases. It sleeps from about five to thirty feet above the ground. Females give birth to one young about every six months.

The Silky Anteater has several adaptations for its treetop existence. Its prehensile tail and an opposable pad on the feet opposite the claws allow the anteater to cling to very slender branches while feeding. If approached and threatened, it raises its body upward, bracing itself with its tail, and raises its forelegs up so that if a predator comes close it can rapidly slice downward with its sharp claws.

Silky Anteaters inhabit moist and wet tropical lowlands and middle elevations on the Caribbean slope and in the southwestern Pacific lowlands and middle elevations. Although rarely seen, it is likely more abundant than observations suggest because of the thick arboreal cover that it occupies. There is one habitat, however, that offers a reasonable chance of spotting this anteater, the mangrove forest. The author has observed the Silky Anteater three times in the Damas Island mangrove forests while on boat tours out of Quepos.

Cyclopes didactylus

Costa Rican names: *Ceibita; tapacara; serafín del platanar.*

2/23 trips; 3 sightings.

Total length: 11.5–18 inches, including 6.7–9.5 inch tail.

Weight: 5.5–9.7 ounces (155–275 grams).

Range: Southern Mexico to Brazil.

Elevational range: Sea level to 5,000 feet.

Silky Anteater, looking like a golden tennis ball as it sleeps

THREE-TOED SLOTH (BROWN-THROATED THREE-TOED SLOTH)

Bradypus variegatus
Costa Rican name: *Perezoso de tres dedos.*
16/23 trips; 38 sightings.
Total length: 22.4–26.0 inches, including 2.6–2.8-inch tail.
Weight: 5 pounds 1 ounce–12 pounds 2 ounces (2.3–5.5 kilograms).
Range: Honduras to northern Argentina.
Elevational range: Sea level to 1,800 feet.

The Three-toed Sloth is one of the most distinctive and abundant rainforest mammals. Each foreleg has three long, curved claws that help grasp branches as it climbs. It has coarse pale brown to grayish hair that often has a greenish tinge caused by algae. The face is white with a dark stripe extending into each eye. There is often a light patch on the throat and chest. In a way, sloths are cold-blooded. Their body temperature drops as much as twelve degrees at night to conserve energy. Each morning they climb to treetops to warm up in the sun.

There is a myth that sloths eat only *Cecropia* leaves, but they actually eat leaves from at least ninety-six trees and vines. Since sloths are often in dense foliage, they are not usually seen except in sparsely branched trees like *Cecropia*. Each sloth eats leaves from a preferred combination of plants, selected from about forty tree species. Bacteria in the sloth's stomach are adapted to digesting those leaves, much like bacteria that digest grass in the stomach of a cow. The combination of tree species needed is different for each sloth, so they can live at higher densities than if they all depended on the same plant. Densities may be as great as three sloths per acre but are usually about one per acre.

A sloth reaches sexual maturity at three years of age and may live twenty to thirty years. An infant clings to the mother for its first six months of life. During that time the young learns which leaves to eat from the female. The mother chews the leaves and passes them to the young so that the bacteria necessary for digestion are transferred from the mother to the young. Then the female leaves the young sloth alone in her territory for six months so it can learn to survive on its own. When the young sloth reaches one year of age, the female returns and forces the young one to find its own territory.

Three-toed Sloths are most

Three-toed Sloth

Three-toed Sloth male, with patch of orange and brown fur on the back

abundant in the Caribbean lowlands, especially in Cahuita NP and the nearby Cahuita vicinity. They can be seen along canals of Tortuguero NP, in *Cecropia* trees just east of Guapiles along the main highway from Guapiles to Limón, in the Central Park at Limón, and along the highway from Limón to Cahuita. They are also found at La Selva Biological Field Station and on the Pacific slope near Quepos, Manuel Antonio NP (along Perezoso Trail), Corcovado NP, and Tiskita Jungle Lodge.

Three-toed Sloths, female climbing with young on belly

Young Three-toed Sloth

TWO-TOED SLOTH

Choloepus hoffmanni
Costa Rican name: *Perezoso de dos dedos.*
13/23 trips; 28 sightings.
Total length: 21.3–27.6 inches; no tail.
Weight: 8 pounds 13 ounces–17 pounds 7 ounces (4–8 kilograms).
Range: Nicaragua to Colombia.
Elevational range: Sea level to 9,200 feet.

The Two-toed Sloth is distinguished from the Three-toed Sloth by its long yellowish-brown to gray fur that gives the appearance of a cheap, faded blond wig. The short, somewhat piglike snout is also different from the less pronounced nose of the Three-toed Sloth. Each foreleg has two long, curved claws. The Three-toed Sloth is found only in lowlands, but this sloth is also found up to middle elevations and cloud forests.

Two-toed Sloths are more nocturnal and include more fruit in their diet than Three-toed Sloths. They can be very difficult to see when curled asleep in the treetops. They occur at a lower density than Three-toed Sloths—about one animal per five to seven acres.

Both sloths together account for about 70 percent of the combined weight of all rainforest mammals. Their densities greatly exceed those of monkeys. In one square mile of forest, there might be more than 1,900 sloths and 190 howler monkeys. Howler monkeys, however, are more conspicuous because they are so noisy and active.

Once a week each sloth descends to the base of the tree on which it has been feeding—and defecates. This fertilizes the tree with nutrients from the leaves that were eaten by the sloth. As this happens, sloth moths fly from the hair of the sloth and lay eggs on the feces. After hatching, sloth moth larvae eat sloth droppings as food. After the larvae develop into adult moths, they return to the forest canopy to locate a sloth. These moths feed on algae that grows within grooves of the sloth's hair. These insects are among many moths, beetles, mites, and ticks that live only on the

Two-toed Sloth adult, on weekly trip to ground to defecate at Limón Central Park

bodies of sloths. As many as 900 insects may live on a single sloth.

The best place to see Two-toed Sloths is in the Central Park at Limón. There has been a population of sloths in this urban park for at least forty years. This sloth can also be observed in trees along the highway from Limón south to Puerto Viejo, in Cahuita NP, and along the canals of Tortuguero NP. They occur in the moist and wet forests of the Osa Peninsula, including Corcovado NP and Tiskita Jungle Lodge. Sometimes they can be spotted in middle-elevation cloud forests of the Monteverde Cloud Forest Reserve, at Monteverde's Ecological Farm, in the central city park at Orotina, in La Selva, in Carara NP, at Tiskita Jungle Lodge, and in the Turrialba region.

Two-toed Sloth adult, climbing a tree, showing its two claws

Two-toed Sloths, female with young. Note sloth moths on the fur.

NINE-BANDED ARMADILLO

Dasypus novemcinctus
Costa Rican names: *Cusuco; armado.*
7/23 trips; 7 sightings.
Total length: 27.0–40.3 inches, including 11.4–17.7-inch tail.
Weight: 7 pounds 1 ounce–15 pounds 7 ounces (3.2–7 kilograms).
Range: Southwestern Missouri to northeastern Argentina.
Elevational range: Sea level to at least 8,000 feet.

The armored appearance, pointed snout, and long, bare tail of the armadillo leave little doubt about its identity. This mammal evolved in South America and has steadily spread northward through Central America and into the United States. Armadillos are in an order of mammals called Edentata—meaning mammals without teeth. Actually, Nine-banded Armadillos have tiny peglike teeth that help in eating ants, termites, caterpillars, beetles, slugs, earthworms, centipedes, and fruits.

The bony plates on the body provide protection from most predators except jaguars, pumas, large dogs, and coyotes. The bands around the central portion of the body provide flexibility (the number of bands may vary from seven to ten, but the usual number is nine). The shell of the armadillo suggests that it is slow-moving like a turtle, but it is as agile as a rabbit and can rapidly escape from a predator by running away, by leaping straight upward, or by quickly digging a hole with its well-developed claws.

Nine-banded Armadillos reach sexual maturity at one year of age. After mating, the fertilized egg undergoes delayed implantation, whereby embryo development is delayed for up to 170 days. Then the egg divides into four separate and identical eggs. This results in identical quadruplets, which are born about 70 days later. Armadillos have an average life span of four years.

Nine-banded Armadillo

Armadillos live in the dry forests of Guanacaste, lowland moist and wet forests of the Caribbean and southern Pacific regions, middle elevations, and cloud forests. They can be common around farms, yards, and gardens as well as pristine forests. The best time to look for armadillos is at night with a flashlight as they explore yards and gardens. During the daytime they usually sleep in underground burrows, but they can occasionally be encountered early or late in the day. Sightings have been made on the grounds of La Pacífica, Palo Verde NP, Sueño Azul, and at CATIE near Turrialba.

SQUIRREL FAMILY *(Sciuridae)*

Variegated Squirrel, rusty color phase

VARIEGATED SQUIRREL

The Variegated Squirrel is the largest squirrel in Costa Rica. Highly variable color patterns range from coal black and grizzled black to rufous and light gray. The sides are frequently silvery gray above and rufous below. The belly is white to cinnamon, and the tail may be black above with rufous, tawny, and white hairs below. This squirrel is in the same genus as the Fox Squirrel and the Gray Squirrel of the United States.

Its diet comprises acorns, nuts, fruits, buds, green plant parts, insects, bird eggs, and small reptiles. Breeding occurs primarily from January through June. Litters of four to eight young are born after a gestation period of forty-five days. The young are raised in tree cavities or leaf nests.

A resident throughout the country except for highlands and the Pacific lowlands south of Manuel Antonio NP, this squirrel is most conspicuous and abundant in Guanacaste. Active during the day, the Variegated Squirrel may be encountered in tropical dry forests of Palo Verde, Guanacaste, and Santa Rosa NPs and in farm groves, riparian forests, and scattered woodlots. Other sightings may be expected in the Caribbean lowlands from La Selva southeast to Cahuita and on the Pacific coast at Carara NP, Manuel Antonio NP, and in the Central Plateau, including the grounds of the Parque Bolívar Zoo.

Sciurus variegatoides

Costa Rican name: *Ardilla.*

18/23 trips; 51 sightings.

Total length: 20.1–22.0 inches, including 9.4–12.0-inch tail.

Weight: 1 pound 0.1 ounce–2 pounds 0.1 ounce (450–909 grams).

Range: Chiapas, Mexico, to central Panama.

Elevational range: Sea level to 5,700 feet.

Variegated Squirrel, gray, white, and rust color phase

RED-TAILED SQUIRREL (NEOTROPICAL RED SQUIRREL)

Sciurus granatensis

Costa Rican names: *Ardilla roja; ardilla chisa; chiza.*

15/23 trips; 34 sightings.

Total length: 13.0–20.5 inches, including 5.5–11.0-inch tail.

Weight: 8.0 ounces–1 pound 2.3 ounces (228–520 grams).

Range: Northwestern Costa Rica to Ecuador, Colombia, and Venezuela.

Elevational range: Sea level to 9,000 feet.

The Red-tailed Squirrel is smaller, darker, and more uniform in color than the Variegated Squirrel. It ranges from black to rusty brown above and light yellowish to rufous-brown below. Although scarce in Guanacaste, it is more common than the Variegated Squirrel in Costa Rica's moist and wet lowland and middle-elevation forests. It is active throughout the day.

Where people put out bananas to feed birds (for example, at Wilson Botanical Garden), this squirrel regularly visits the feeders. Large fruits, palm fruits, legumes, mushrooms, young leaves, flowers, and tree bark make up much of the diet. Red-tailed Squirrels forage on the ground and in trees.

This squirrel produces two or three litters per year, and each litter averages two young. Mating activity is especially apparent during the dry season. Four to eight males pursue a female through the treetops for several hours prior to mating. Young are born after forty-four days and are cared for by the female for eight to ten weeks. Red-tailed Squirrels occur at densities up to four per acre. The average home range is about 1.5 acres for females and 3.7 acres for males.

Red-tailed Squirrel, carrying an avocado

Alfaro's Pygmy Squirrel

Microsciurus alfari

Costa Rican name: *Ardilla enana.* 10/23 trips; 12 sightings.

Total length: 7.4–10.8 inches, including 3.1–5.1-inch tail.

Weight: 2.5–3.7 ounces (72–105 grams).

Range: Southern Nicaragua to northwestern Colombia

Elevational range: Sea level to 7,800 feet.

ALFARO'S PYGMY SQUIRREL

Alfaro's Pygmy Squirrel is tiny and dark, with a body only four to five inches long. An inhabitant of rainforests from lowlands to montane levels, it may be seen foraging for seeds, palm fruits, or nuts on the ground and in upper levels of the rainforest canopy.

There is also a history lesson in this squirrel's name. Anastasio Alfaro was the first director of Costa Rica's National Museum. A pioneer naturalist, he was appointed to that position in 1887 at the age of 22.

Alfaro's Pygmy Squirrel is a slender squirrel that is dark gray above and lighter gray to yellowish below. The ears are small, and the tail is slender, rather than bushy like the tail of most squirrels.

Foods include fruits of *Scheelea* palms and the almendro tree (*Dipteryx panamensis*). This squirrel also consumes the sap of rainforest trees. It clings to the trunk of a large *Quararibea costaricensis* or *Inga* tree as it gnaws small holes in the bark and laps the sap.

Alfaro's Pygmy Squirrel appears to fill an ecological niche as a small sap-lapping arboreal mammal, comparable to the niche occupied by the Common Pygmy Marmoset in lowland rainforests of Ecuador, Peru, and Brazil. The pygmy marmoset is the smallest primate in the world, but it looks more like a squirrel than a monkey. Its body measures only 117 to 152 millimeters, a little smaller than Alfaro's Pygmy Squirrel, which measures 123 to 159 millimeters. The squirrel, however, is lighter, weighing in at only 72–105 grams to the marmoset's 85–141 grams. The pygmy marmoset uses its well-developed incisor teeth to open sap holes on large tree trunks. It clings to the side of a tree much the same as Alfaro's Pygmy Squirrels, even feeding on some of the same genera of trees, like *Inga* trees. Alfaro's Pygmy Squirrels have been poorly studied, and few details are known about their life history and reproduction.

These pygmy squirrels have been encountered at middle elevations of the Caribbean slope in the Monteverde Cloud Forest Reserve and at Rancho Naturalista. They are more commonly encountered in the moist and wet forests of the southern Pacific lowlands, including the grounds of Hotel Villa Lapas, La Cusinga Lodge, Tapanti NP, Esquinas Rainforest Lodge, and on the grounds of Corcovado Lodge Tent Camp on the Osa Peninsula.

MEXICAN HAIRY PORCUPINE (PREHENSILE-TAILED PORCUPINE)

Coendu mexicanus

Costa Rican name: *Puerco espín.*

4/23 trips; 5 sightings.

Total length: 31.3–39.4 inches, including 17.1–19.1-inch tail.

Weight: 3 pounds 1 ounce–5 pounds 12 ounces (1.4–2.6 kilograms).

Range: Mexico to Panama.

Elevational range: Sea level to 9,800 feet.

This tropical porcupine is an agile and intelligent mammal that shares a feature with opossums, a prehensile tail. Although the North American Porcupine has a short tail thickly covered with quills, this porcupine has a long prehensile tail that is bare at the tip. It can hang by its tail from tree branches as it reaches for fruits, blossoms, young leaves, seeds, and nuts. While eating, this porcupine holds fruits and seeds in its paws like a squirrel. One interesting adaptation is that when this porcupine is in the treetops, the fur and spines lie parallel to the contours of the body. When the porcupine is on the ground, it fluffs out the fur and spines so they are perpendicular to the contours of the body. This makes the animal look larger and more intimidating to potential predators (Winnie Hallwachs, personal communication 2001).

Although widespread, this porcupine is seldom seen. In Guanacaste it is more conspicuous in the deciduous trees because of the sparse foliage. It occurs up to middle and high elevations, as in the Monteverde Cloud Forest Reserve, and up to 8,500 feet in Cerro de la Muerte. It lives in dry, moist, and wet forests of both Caribbean and Pacific slopes. With the aid of spotlights, these porcupines are sometimes seen at night as they cross roads or feed in the treetops in Cerro de la Muerte. They can be encountered on night drives along the road that descends from kilometer 80 on the Pan-American Highway to Savegre Mountain Lodge. These porcupines sleep in hollow trees during the day.

Mexican Hairy Porcupine in the Talamanca Mountains

Mexican Hairy Porcupine

Mexican Hairy Porcupine, climbing a tree

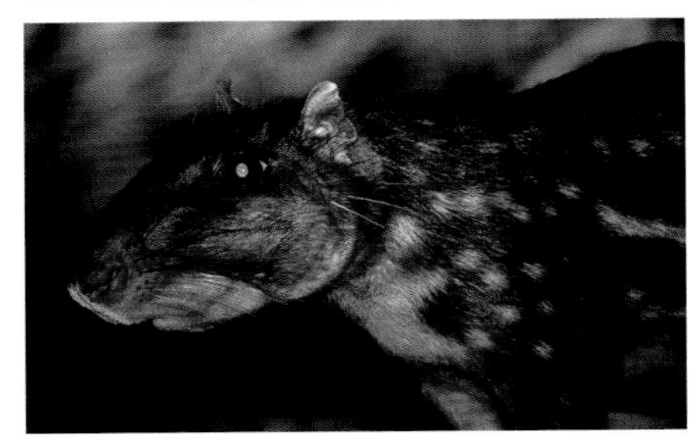

Tepescuintle

TEPESCUINTLE (PACA)

Cuniculus paca (formerly Agouti paca)

Costa Rican name: *Tepezcuintle.*

2/23 trips; 4 sightings.

Total length: 29.1 inches, including 0.9-inch tail.

Weight: 8 pounds 13 ounces–26 pounds 7 ounces (5.8–12 kilograms).

Range: Southern Mexico to southeastern Brazil.

Elevational range: Sea level to 6,600 feet.

About the size of a cocker spaniel, the tepescuintle, or paca, is the largest rodent in Costa Rica. In the past, it was considered a member of the Dasyproctidae, related to guinea pigs and agoutis. The rows of whitish spots along the sides provide camouflage and suggest the pattern of a White-tailed Deer fawn. Enlarged, hollow cheek areas on the male's skull serve as resonating chambers to amplify the loud and ominous roar the animal makes when fighting or defending itself. Female pacas do not have resonating chambers. Intensive hunting pressure has made the tepescuintle rare in many areas because it is highly regarded for the quality of its meat.

The tepescuintle is nocturnal and solitary, except for females accompanied by young. A female has only one young in May or June after a gestation period of 115 days. It spends daytime hours in a shallow burrow under tree roots or in a hollow log—often near water. At night it searches for fallen fruits, seeds, leaves, stems, and roots. It will also eat fallen mangoes or avocados in orchards and may visit gardens to eat corn, watermelons, or squash. If pursued, a tepescuintle has two escape strategies: it will either run a short distance and stand motionless for up to forty-five minutes, or it will leap into a marsh or stream and submerge for many minutes with only the eyes and nostrils exposed.

Habitat ranges from riparian forests in Guanacaste to moist and wet rainforests of the Caribbean and southern Pacific regions. This species can survive in forest fragments, farmland, and edges of cities where thickets provide adequate hiding places.

The best place in Costa Rica to see tepescuintles is at Bosque de Paz, where they come into the courtyard after dark to eat cracked corn from troughs that are placed for them by the owners of the lodge. Visitors who sit quietly on the lawn can observe this rare and shy mammal at a distance of less than twenty-five feet.

Tepescuintle adult

AGOUTI

Dasyprocta punctata
Costa Rican names: *Guatusa; guatuza; cherenga.*
14/23 trips; 26 sightings.
Total length: 16.3–24.4 inches, including 0.4–1.0-inch tail.
Weight: 5 pounds–8 pounds 13 ounces (2.27–4.00 kilograms).
Range: Mexico to northern Argentina.
Elevational range: Sea level to 7,900 feet.

The agouti is a medium-sized mammal that looks like a short-eared, chestnut-brown rabbit with a tiny hairless tail. It gives the appearance of walking on its tiptoes and sits upright to eat seeds in its paws like a squirrel. The agouti is prey for jaguars, pumas, ocelots, margays, tayras, coatis, boa constrictors, and large raptors. If approached, an agouti will erect the long hairs on its rump and thump its hind feet on the ground. If that is not successful in deterring a predator, it will race through the underbrush while making high-pitched barks.

Habitat ranges from dry forests in Guanacaste to moist and wet forests of the Caribbean and Pacific lowlands. In areas protected from hunting, the agouti is active in daytime. With constant hunting pressure, it adapts to a nocturnal existence. Hollow logs or burrows are used as dens.

The agouti is an important disperser of forest seeds. It collects and buries seeds from such plants as almendro (*Dipteryx panamensis*), pejivalle (*Bactris gasipes*), and caobilla (*Carapa guianensis*) in times of seed abundance and retrieves them in times of food scarcity. Seeds not eaten may eventually germinate and grow.

A pair of agoutis occupies a home range of 5 to 9.6 acres. The male has an unusual habit of spraying the female with urine during its courtship ritual to attract her attention. The subsequent gestation period is forty-four days, and the litter size is usually one or two. After the young are born, the female places them in tiny burrows that are too small for her to enter. This provides protection from predators. She calls them out to nurse.

One of the best places to see agoutis during the day is at La Selva Biological Field Station. Look for them on the lawn among the cabins across the Stone suspension bridge from the headquarters and along the Tres Ríos trail. They can also be seen crossing trails at Carara, Santa Rosa, and Manuel Antonio NPs; at Corcovado Lodge Tent Camp; at Los Cusingos; and on the grounds of the Wilson Botanical Garden.

Agouti adult female, nursing young

Agouti adult

Coyote adult

COYOTE

The wily coyote has been a successful and adaptable resident in Costa Rica. It occurs from frigid northern environments in Alaska south to the tropical dry forests of Guanacaste and the paramo of Cerro de la Muerte. The range of the coyote was originally limited to Guanacaste in 1960, but in the last four decades it has expanded its range through the Talamanca Mountains and into western Panama.

In Guanacaste, coyotes prey on iguanas, ctenosaurs, peccaries, deer, small rodents, snakes, tinamous, and ducks. In the highlands, prey species include rabbits, armadillos, brocket deer, and Band-tailed Pigeons. Carrion is also eaten, including dead livestock on farms and ranches.

Coyotes of the tropical dry forest tend to be smaller than heavy-bodied coyotes of northern climates. This is an example of Bergmann's rule, which states that the body size of mammals, within the same species, is larger in more northern latitudes in order to help preserve body heat in colder weather. In tropical environments, smaller body size helps the animal lose excessive heat more quickly. Bergmann's rule applies to larger Costa Rican mammals like the puma, White-tailed Deer, Gray Fox, Northern Raccoon, and coyote.

Coyotes may be observed in packs and may be heard at night as they howl. They are most active at dusk, at night, and early in the morning. A pair will typically have a den where the female gives birth to litters ranging from two to twelve young after a gestation period of two months.

Coyotes are most likely to be observed by driving roads at night on ranches and in areas of mixed forest and grassland in Guanacaste.

Canis latrans

Costa Rican name: *Coyote.*

Total length: 39.4–61.0 inches, including 11.8–15.7-inch tail.

Weight: 18–33 pounds (8.2–15.0 kilograms).

Range: Alaska to western Panama.

Elevational range: Sea level to 9,000 feet.

GRAY FOX

Urocyon cinereoargenteus
Costa Rican name: *Zorro gris.*
1/24 trips; 1 sighting.
Total length: 33.8–44.0 inches, including 10.9–15.7-inch tail.
Weight: 3 pounds 14 ounces–7 pounds 11 ounces (1.8–3.5 kilograms).
Range: Southern Canada to Venezuela.
Elevational range: Sea level to 7,800 feet.

The Gray Fox is an attractive member of the dog family, but it acts more like a cat in the way that it hunts and moves. It can even climb trees. This is possible because the Gray Fox can rotate the wrists of its forelegs to grasp a tree trunk as it climbs, and its claws are partially retractable to facilitate climbing. This fox has short legs, and its head and muzzle are proportionately smaller than a coyote's. The Gray Foxes of Central America are smaller than those of northern temperate regions, another example of Bergmann's rule.

The Gray Fox is found in highland and lowland habitats. It is more common in Guanacaste and in the Monteverde region, where it is mainly nocturnal and seldom seen. It is not found in the Caribbean lowlands. An omnivorous species, this fox eats lizards, rodents, birds, and insects. Females give birth to one litter per year after a gestation period of two months. A litter will have three to seven young. The den may be in a tunnel, in a rock pile, or under a brush pile.

Gray Foxes may sometimes be seen in the Monteverde community and during night drives in the Guanacaste region.

Gray Fox adult

TAYRA

Eira barbara

Costa Rican names: *Tolumuco; gato de monte; cholumuco; tejón; cabeza de viejo.*

4/23 trips; 5 sightings.

Total length: 38.5–45.2 inches, including 15.0–18.5-inch tail.

Weight: 8 pounds 13 ounces–13 pounds 4 ounces (4–6 kilograms).

Range: Central Mexico to northern Argentina.

Elevational range: Sea level to 7,400 feet.

The tayra, about the size of a fox, is dark brown to black. It is typically glimpsed as it runs up and down tree trunks and along tree branches in search of prey. The tail is long and bushy. The head may be black, dark brown, or light gray contrasting with a dark brown body. This gives rise to the Spanish name *cabeza de viejo,* meaning "old man's head." Most tayras in Costa Rica have a dark head. The tayra appears to fill an ecological niche as a predator similar to that of the fisher, an arboreal fox-sized member of the weasel family in the boreal forests of the northern United States and Canada.

The tayra is active throughout the day. Its relentless hunting behavior occurs on the ground and in trees. Prey includes just about anything smaller than a White-tailed Deer—figs, fruits, domestic poultry, sloths, squirrels, agoutis, tepescuintles, mice, rats, lizards, wild birds, and eggs. This mammal is usually solitary. Occasionally tayras are encountered as pairs or females with young. They may even hunt in groups of up to twenty animals.

Habitats occupied by the tayra range from dry and riparian forests of Guanacaste to wet lowland forests of the Caribbean and southern Pacific slopes. The species occurs at higher elevations throughout the mountains of Costa Rica. Habitat includes farms, plantations, and undisturbed forest.

The tayra makes its den in a hollow tree, hollow log, or sheltered burrow. It has one to three young after a gestation period of sixty-five to seventy days. When the young are two months old, the female teaches them to hunt. The tayra may be seen along the canals of Tortuguero NP, in the forests of La Selva Biological Field Station, and in Santa Rosa and Guanacaste NPs.

Tayra adult. Most individuals in Costa Rica are black.

CENTRAL AMERICAN RIVER OTTER (NEOTROPICAL RIVER OTTER)

Lutra longicaudis

Costa Rican name: *Perro de agua.*

7/23 trips; 9 sightings.

Total length: 36.8–50.1 inches, including 14.6–18.7-inch tail.

Weight: 11 pounds–20 pounds 14 ounces (5.0–9.5 kilograms).

Range: Mexico to Brazil.

Elevational range: Sea level to 9,000 feet.

This river otter is a member of the weasel family, but it acts like a dog. Its Spanish name, *perro de agua,* means "dog of the water," referring to the barking sounds made by an otter.

Central American River Otters are very adaptable species, found from tropical lowlands to montane levels. They occupy ponds, rivers, and mountain streams with clear water, where they live on a diet of fish, crabs, shrimp, crayfish, and the occasional frog or bird. They learn to capture fish and to turn over rocks on the bottom of a stream to find fish and shrimp concealed under the rocks.

Since otters are diurnal, they may occasionally be spotted from boats along tropical streams and canals up to high elevations. They maintain burrows along riverbanks, where they sleep and raise their young. From one to five pups are born after a gestation period of two months. They apparently do not go through a period of delayed implantation, as do northern river otters.

Otters are uncommon because they were once killed for their pelts and extirpated. They may have benefited at higher elevations from the introduction of exotic rainbow trout, which provide an additional source of food in mountain streams.

River otters can be seen in the canals in Tortuguero NP and at Esquinas Rainforest Lodge in southwestern Costa Rica.

Central American River Otter in Tortuguero National Park

Kinkajou adult male. Note the prehensile tail.

KINKAJOU

The kinkajou is sometimes called "honey bear," because of its habit of inserting its long tongue into bee hives so it can lick the honey. Although it is nocturnal and seldom seen, its appealing face, soft gray fur, and prehensile tail have made it a well-known tropical mammal.

Habitat varies from riparian forests of Guanacaste to lowland wet forests in the Caribbean and southern Pacific slopes, cloud forests, and montane forests. It lives in treetop environments where its diet comprises insects, grubs, small mammals, birds, bird eggs, honey, bananas, and wild fruits. Fruits eaten include guava (*Psidium*), inga (*Inga coruscans*), jobo (*Spondias mombin*), palm (*Welfia georgia*), and nectar of balsa (*Ochroma pyramidale*) flowers. The kinkajou is ecologically important because it disperses fruit seeds in its droppings. When a tree is full of fruits, seven or eight kinkajous may gather at night to feed. They make their presence known with noisy barks, whistles, and shrill screams that can be heard up to a mile away. In good habitat, this mammal can reach a density of three to four animals per acre.

The kinkajou usually has one young after a gestation period of ninety days. The den is typically in a hollow tree. In the past it has been hunted to capture the young ones as pets. It makes a very poor pet, however, since it sleeps during the day and makes noise all night. The nocturnal kinkajou is rarely seen except during night hikes using spotlights. This species can be seen along the canals of Tortuguero NP and at La Virgen del Socorro, La Selva Biological Field Station, the Ecological Farm at Monteverde, La Pacífica, and the Sirena Biological Station.

Potos flavus

Costa Rican names: *Marta*; *martilla*; *oso mielero*.

8/23 trips; 13 sightings.

Total length: 34.8–41.3 inches, including 18.1–21.1-inch tail.

Weight: 4 pounds 6 ounces–10 pounds 2 ounces (2.0–4.6 kilograms).

Range: Southern Mexico to southern Brazil.

Elevational range: Sea level to 6,800 feet.

OLINGO

Bassaricyon gabbii
Costa Rican name: *Olingo; martilla.*
0/23 trips; 0 sightings.
Length: 29.1–35.4 inches, including 15.0–18.9-inch tail.
Weight: 2 pounds 6.7 ounces–3 pounds 1.6 ounces (1.1–1.4 kilograms).
Range: Central Nicaragua to western Colombia.
Elevational range: Sea level to 1,700 feet.

One of the rarest members of the raccoon family in Costa Rica is a shy, nocturnal mammal called the olingo. It resembles a slender kinkajou, but there are some interesting differences. The heavier kinkajou has a long prehensile tail, which it can wrap around a tree branch for security as it feeds. The lighter and more agile olingo does not have a prehensile tail, but like a squirrel monkey, its long and furry tail helps the olingo maintain balance when jumping from one branch to another.

The olingo's snout is more pointed than the kinkajou's. The olingo is also much more athletic than the kinkajou, making long leaps, up to about ten feet, from one branch to another. The range is limited from Nicaragua to western Ecuador, and within that range the olingo lives mainly in mature moist and wet forests at middle-elevation cloud forest levels.

Primary foods include fruits like figs, the nectar of flowers from plants like balsa, invertebrates, and small birds and animals. When visiting balsa and kapok trees to drink the nectar, olingos appear to pollinate the flowers as well. When food is abundant in a fruiting tree, olingos, kinkajous, and Common Opossums may all feed together. Olingos feed at night and sleep in tree cavities during the day.

This rare mammal reaches sexual maturity at the age of two years. Young are born in the dry season (January through March) after a gestation period of about two and a half months. Females give birth to a single young. This species may live up to seventeen years.

A sighting of an olingo is a rare event, but these animals appear to be more common in the Monteverde community than elsewhere in Costa Rica. For many years an olingo at Monteverde has visited the Hummingbird Gallery each afternoon to spill sugar water from the hummingbird feeders so it can lap up the sweet mixture. The olingo shown here was photographed at the Hummingbird Gallery by Robert Djupstrom of White Bear Lake, Minnesota.

Olingo adult. Note that the tail is not prehensile. Photo © Robert Djupstrom

White-nosed Coati adult male

Nasua narica (formerly Nasua nasua)

Costa Rican names: *Pizote; coatimundi.*

13/23 trips; 24 sightings.

Total length: 33.5–54.3 inches, including 16.5–26.8-inch tail.

Weight: 6 pounds 10 ounces–15 pounds 7 ounces (3–7 kilograms).

Range: Texas, New Mexico, and Arizona to northwestern Colombia.

Elevational range: Sea level to 6,900 feet.

WHITE-NOSED COATI (COATIMUNDI)

The White-nosed Coati, also known as the coatimundi or coati, is the largest and most conspicuous member of the raccoon family in Costa Rica. A social mammal, it is active throughout the day and is usually seen in family groups of fifteen to twenty. Groups may occasionally exceed thirty individuals. These groups consist of young males, females, and young. Old males are typically loners. Distinctive features are the long, dark, furry tail, which is held upright as it forages on the ground, and the highly flexible nose, which is used to sniff for food.

The coati is at home on the ground or in trees. Its omnivorous diet includes fruit, nuts, figs, insects, lizards, small mammals, birds, bird eggs, snails, worms, crabs, insect larvae, carrion, turtle eggs, and snakes. At night a group of coatis sleeps in the treetops. If the group is attacked, or perhaps shot at, the entire troop drops out of the treetop in a bewildering array of falling bodies that confuses the predator. Then all the coatis race off to safety. Coatis can fight fiercely as a group, so they have few predators. Jaguars, pumas, tayras, hawk-eagles, and boa constrictors will occasionally prey on a coatimundi. White-throated Capuchins will sometimes prey on baby coatis in their nests.

A band of coatis led by adult females may occupy a home range of 75 to 125 acres. During the breeding season, a solitary male will briefly join a band of coatis for mating. From two to four young are born in a leafy treetop nest platform after a gestation of seventy to seventy-seven days. Young stay in the nest for about five weeks. In Guanacaste they may reproduce twice each year.

This mammal is found in habitats ranging from dry forests in Guanacaste to moist and wet forests of the Caribbean and southern Pacific slopes. The coati is most conspicuous, however, in Guanacaste, where family groups are frequently encountered crossing roads and trails in Palo Verde, Guanacaste, and Santa Rosa NPs and in Lomas Barbudal BR. This species is also encountered at Hacienda Solimar, Corcovado Lodge Tent Camp, and Wilson Botanical Garden and in the Monteverde vicinity at the Ecological Farm, Monteverde Lodge, Hummingbird Gallery, and the Cloud Forest Reserve headquarters.

White-nosed Coati

CRAB-EATING RACCOON

The range of this South American member of the raccoon family extends primarily into the southwestern Pacific lowlands of Costa Rica, overlapping with the Northern Raccoon. There may be some confusion as to which raccoon is which in that region. The legs of the Crab-eating Raccoon are dark brown to black, where those of the Northern Raccoon are light brown to gray. Other characteristics overlap; for instance, both raccoons can be found along muddy shores of mangrove lagoons where the Crab-eating Raccoon would be expected, and Crab-eating Raccoons can be found inland where Northern Raccoons would be expected. Other distinguishing characteristics require examination of a dead specimen: the fur on top of the shoulders points toward the head on a Crab-eating Raccoon and backward on a Northern Raccoon. The mask is shorter on the Crab-eating Raccoon, and it also has no underfur, only coarse guard hairs, so it looks less fluffy than the Northern Raccoon.

This uncommon raccoon is less omnivorous than the Northern Raccoon. It eats crabs, fish, snails, fruits, frogs, and other invertebrates. It is typically found in coastal areas, mangrove forests, and riparian areas where aquatic species are abundant. The Crab-eating Raccoon is nocturnal and sleeps in hollow trees during the day. This raccoon could be expected in mangrove lagoons in the Quepos area and on the Osa Peninsula, and in coastal areas east of the Golfo Dulce near Panama. The Crab-eating Raccoon shown here was photographed at night at Hato El Cedral on the llanos of Venezuela.

Procyon cancrivorus

Costa Rican names: *Mapache; mapachín.*

2/23 trips; 2 sightings.

Total length: 31.4–45.0 inches, including 9.8–15.4-inch tail.

Weight: 6 pounds 9.6 ounces–15 pounds 6.4 ounces (3–7 kilograms).

Range: Pacific coast of Costa Rica to northern Argentina.

Elevational range: Sea level to 4,000 feet.

Crab-eating Raccoon (photographed on the llanos of Venezuela, at Hato El Cedral)

NORTHERN RACCOON

Procyon lotor

Costa Rican names: *Mapache; mapachín.*

6/23 trips; 7 sightings.

Total length: 25.6–37.4 inches, including 10.1–13.0-inch tail.

Weight: 7 pounds 4.2 ounces–17 pounds 2.6 ounces (3.0–7.8 kilograms.

Range: Southern Canada to central Panama.

Elevational range: Sea level to 8,400 feet.

Costa Rica and Panama are the only two countries in the Americas with two raccoon species, the Northern and the Crab-eating Raccoons. Like the coyote, the Northern Raccoon is found from Canada to Panama. Also like the coyote, it conforms to Bergmann's rule, which states that within a given species, the northern animals are larger in order to conserve body heat in cold weather, and the tropical cousins are smaller in order to facilitate cooling in hot weather. The Northern Raccoon is more light brown to grayish than the Crab-eating Raccoon, with pale feet rather than dark brown to black feet and legs. The mask extends farther back on the face behind the eyes, and Northern Raccoons have a layer of underfur that gives them a furrier look than the Crab-eating Raccoon.

Raccoons are omnivorous, nocturnal, and arboreal. They usually sleep in hollow tree cavities during the day and come out at night to feed on crayfish, crabs, frogs, and fruits. In settled areas they will eat field corn and assorted human garbage. In mangrove areas they appear in the daytime along shores at low tide to forage.

Northern Raccoons are found throughout much of Costa Rica, but most sightings are in forested coastal regions, in mangroves, in wetlands along the Caribbean and Pacific coasts, and along rivers in inland areas from premontane and montane levels. Sightings can be expected in Tortuguero and Cahuita NPs, La Selva, Selva Verde, and Bosque de Paz on the Caribbean slope. On the Pacific coast, they have been observed at Palo Verde, Santa Rosa, and Carara NPs and in the mangrove lagoons of the Río Abangares estuary in the Gulf of Nicoya. Raccoons can be seen at dusk in the courtyard of Bosque de Paz where they eat cracked corn that has been placed in troughs to feed the coatis, raccoons, agoutis, and tepescuintles.

Northern Raccoon adult in Carara National Park

Northern Raccoon adult among the mangroves at La Ensenada Lodge

OCELOT

Leopardus pardalis

Costa Rican names: *Manigordo; ocelote.*

1/23 trips; 1 sighting.

Total length: 38.5–51.2 inches, including 11.0–15.7-inch tail.

Weight: 17 pounds 10 ounces–33 pounds 2 ounces (8,000–15,000 grams).

Range: Southern Texas to northern Argentina.

Elevational range: Sea level to 8,600 feet, with records over 10,000 feet.

Although rarely seen, wild cats evoke a great sense of anticipation and excitement among rainforest travelers who hope to glimpse one. There are six wild cats in Costa Rica: jaguar, puma, jaguarundi, ocelot, margay, and oncilla. Human demand for the spotted pelt of the ocelot has caused populations to be greatly persecuted in the past, but protection as an endangered species is helping to restore its numbers.

The ocelot is a medium-sized cat adapted to riparian and dry forests of Guanacaste, moist and wet forests of the Caribbean and southern Pacific slopes, and montane forests of Braulio Carrillo and La Amistad NPs. It occupies undisturbed forests, second-growth forests, and agricultural areas.

The ocelot is nocturnal, but it also hunts at dawn and dusk. Prey includes spiny rats, agoutis, birds, lizards, opossums, snakes, and amphibians. It usually hunts on the ground, but it can also climb trees. A female ocelot maintains a territory of 250–3,360 acres, which it defends from other females, and the male maintains a territory of 800–4,080 acres, which it defends against other males. The male's territory may include the territories of several females.

Females reach reproductive age at eighteen to twenty-two months, and they may produce young every other year to the age of thirteen. One to two young are born after a gestation period of seventy to eighty days. Cubs stay with the female for up to a year. The ocelot is rarely seen, but it can be encountered, with the aid of spotlights, on night excursions guided by naturalists.

Ocelot, concealed in heavy cover

Ocelot adult

Ocelot head

Margay. The head is proportionately smaller than that of an ocelot.

MARGAY

The beautiful spotted margay is slightly larger than a housecat, and it looks like an arboreal version of an ocelot. The ocelot has a heavier body with a shorter tail and is more terrestrial in its hunting habits. The margay has a long, furry tail (about 70 percent of the length of the head and body) that aids its balance as it moves and jumps among the treetops in search of mammalian and avian prey. Ocelots can climb trees, but that is not their preferred habitat. The margay also has an unusual adaptation for climbing and descending trees with great speed and agility; it can rotate its hind feet sideways, allowing it to grip the bark with its long claws, like a squirrel, as it moves up and down tree trunks.

Leopardus wiedii

Costa Rican names: *Caucel; tigrillo.*

0/23 trips; 0 sightings.

Total length: 33.8–50 inches, including 14.6–21.0-inch tail.

Weight: 5 pounds 11.5 ounces–11 pounds (2.6-5 kilograms).

Range: Border of northern Mexico and Texas to northern Argentina.

Elevational range: Sea level to 9,000 feet

The margay's head and eyes are large, considering its body size. The large eyes are an adaptation for seeing better at night. Nocturnal and arboreal, it sleeps amid treetop vines and in tree cavities during the day. The margay is believed to communicate using eight vocalizations: hissing, spitting, growling, snarling, purring, meowing, barking, and moaning.

The margay is a long-lived species with a low reproductive rate. It reaches sexual maturity at the age of two. The female has only two nipples and usually has only one young after a gestation period of twelve weeks. The young leaves the treetop nest after about five weeks and continues to be nursed by the female to the age of two months. This cat can live to the age of twenty years. Many of these reproductive and longevity details are intriguingly similar to those of the olingo in the raccoon family, which is also a slender treetop carnivore with a long bushy tail.

Prey of the margay includes small arboreal mammals, birds, reptiles, amphibians, and invertebrates as well as some species that are caught on the ground. Among the mammals taken are small opossums, squirrels, mice, rabbits, and agoutis. The birds include songbirds and tinamous. It will also eat some fruits.

Margays are found in mature and older second-growth dry, moist, and wet tropical forests throughout Costa Rica. They are rarely observed, however, and are most likely to be seen along trails at night with the aid of spotlights.

ONCILLA (LITTLE SPOTTED CAT)

The oncilla, also known as the Little Spotted Cat, is an endangered species about the size of a medium to large house cat. Its body proportions are similar to those of the larger ocelot. The tail is proportionately shorter than the tail of a margay. Some individuals are melanistic, with a black background color and distinguishable black spots. This rare, nocturnal cat is more terrestrial than the margay and feeds mainly on small mammals and ground-dwelling birds, including tinamous, wood-quail, sparrows, and finches.

Poorly known and poorly studied, most details of its life history and reproduction are unknown. The gestation period is two and a half months, and a typical litter includes one or two young.

When glimpsed briefly, it would be difficult to distinguish this cat from a margay or small ocelot, but its range is more limited. It inhabits high-elevation forests of the Tilarán and Talamanca Mountains, including Cerro de la Muerte. Birding guide Carlos Gómez has encountered this cat at night calmly sitting along the winding forest road that descends from kilometer 80 of the Pan-American Highway to Savegre Mountain Lodge. It allowed observers to approach quite closely before it ran off.

Leopardus tigrinus
Costa Rican names: *Caucel; tigrillo.*
0/23 trips; 0 sightings.
Length: 26.4–38.9 inches, including 9.6–13.3-inch tail.
Weight: 3 pounds 1.3 ounces–6 pounds 2.6 ounces (1.4–2.8 kilograms).
Range: Costa Rica to Brazil.
Elevational range: Mainly 5,000–10,500 feet in Costa Rica, at lower elevations in South America.

Oncilla, the smallest wild cat of Costa Rica. It is most frequently encountered in the Talamanca Mountains.

JAGUARUNDI

Herpailurus yaguarondi

Costa Rican names: *Yaguarundí; león breñero; león miquero.*

3/23 trips; 3 sightings.

Total length: 39.8–54.7 inches, including a 10.0–19.9-inch tail.

Weight: 8 pounds 13 ounces–19 pounds 13 ounces (4–9 kilograms).

Range: Southern Texas to northern Argentina.

Elevational range: Sea level to 6,600 feet.

One of the most unusual and poorly known wild cats is the jaguarundi. It has a short rounded head like a house cat, a long body, a long slender tail, and short legs. There is a resemblance to the tayra, but the tail is not as bushy. This cat occurs in both reddish and grayish-black color morphs.

The jaguarundi inhabits riparian forests in Guanacaste, lowland moist and wet forests, ranches, farms, and montane forests. Extremely agile and an excellent hunter, the jaguarundi hunts primarily on the ground. It eats quail, wood quail, tinamous, small rodents, armadillos, lizards, and insects. It also pursues prey in trees.

The jaguarundi is the most vocal of Costa Rica's wild cats. It makes thirteen vocalizations, including birdlike chirps. One to four young are born in a hollow tree or den after sixty to seventy-five days of gestation. The average litter size is two. This cat is more active during the day than other cats. Sightings have been made in the vicinity of La Selva, along the roadside near the Río Pijije bridge at Cañas, and at Savegre Mountain Lodge. Most sightings are likely to be a quick glimpse of the animal bounding across a forest road or trail.

Jaguarundi, red color phase

Jaguarundi, gray color phase

PUMA (COUGAR)

The puma resembles a very large, heavy-bodied version of the red-phase jaguarundi. Although subspecies from the American Rockies can reach a length of almost nine feet and weigh over 200 pounds, the pumas of Costa Rica's forests are smaller and more slender and are generally comparable in size and conformation to the subtropical Florida panther. The puma occurs in dry and wet forests from sea level to 11,000 feet but is seldom seen. It preys on White-tailed Deer, brocket deer, tepescuintles, agoutis, armadillos, monkeys, tamanduas, iguanas, raccoons, and, occasionally, cattle and horses. Good populations exist in Tortuguero and Corcovado NPs. Tracks are regularly encountered at the Sirena Biological Station, and they are reported to have been seen more frequently by tourists at that location in recent years.

Puma concolor

Costa Rican names: *Puma; león; león de montaña.*

Length: 59 inches–106 inches.

Weight: 59 pounds 15.2 ounces–69 pounds 15.4 ounces (27.3–31.8 kilograms).

Range: British Columbia to southern Chile and Argentina.

Elevational range: Sea level to 11,000 feet.

Puma. This Costa Rican subspecies is smaller than those of northern latitudes.

Jaguar, melanistic (black) color phase

JAGUAR

Panthera onca

Costa Rican name: *Tigre.*

0/23 trips; 0 sightings.

Total length: 87–120 inches, including 20.5–26.0-inch tail.

Weight: 132–348 pounds (59.9–158.0 kilograms).

Range: Southern Arizona to northern Argentina.

Elevational range: Sea level to 6,600 feet, with some records up to 10,000 feet.

The awesome but endangered jaguar is the largest carnivore in the Americas. Its massive and muscular body enables it to overwhelm prey up to the size of a tapir, and it kills by puncturing the prey's skull with its large canines. A jaguar's body is highlighted by black rosettes on a golden-brown background. Some jaguars are a melanistic black phase in which black rosettes are visible like black velvet against a plain black background. The jaguar has become a symbol of rainforest preservation, because its survival depends on the protection of large tracts of unbroken forest.

The jaguar lives in dry and riparian forests in Santa Rosa and Guanacaste NPs, cloud forests, montane forests, and moist and wet forests in Caribbean and southern Pacific lowlands. The best population in Costa Rica is on the Osa Peninsula in Corcovado NP, where jaguars frequent areas near rivers, lagoons, and beaches.

This cat hunts during the day and night in protected areas where it is not exposed to hunting. It hunts by stalking its prey or by lying in ambush. Prey includes peccaries, White-tailed and Brocket Deer, White-throated Capuchins, tapirs, agoutis, sloths, birds, caimans, fish, iguanas, snakes, and domestic livestock. It is reported to hunt sea turtles at night along the beach near Sirena Biological Station on the Osa Peninsula. Jaguars occupy territories that they defend against other jaguars of the same sex. Males occupy home ranges up to 5,800 acres.

The jaguar matures at three years of age and may live up to eleven years in the wild. Mating may occur at any time of year and is accompanied by very noisy roaring, grunting, and caterwauling. When a female is receptive, a pair may mate up to a hundred times in a day. The gestation period is 90 to 111 days. There are usually one or two cubs in a litter, with some records of three or four. The cubs accompany the female for eighteen to twenty-four months.

Chances of seeing a jaguar in the wild are very low, but one of the best concentrations of these great cats is at Sirena Biological Station in Corcovado NP. Tracks can be observed on the beach, along trails, on sandy banks of the Río Pavo, and along the banks of nearby Río Llorona. A jaguar track is wider than it is long, and forefoot prints are larger than hindfoot prints. A forefoot print can be six inches wide across the toes.

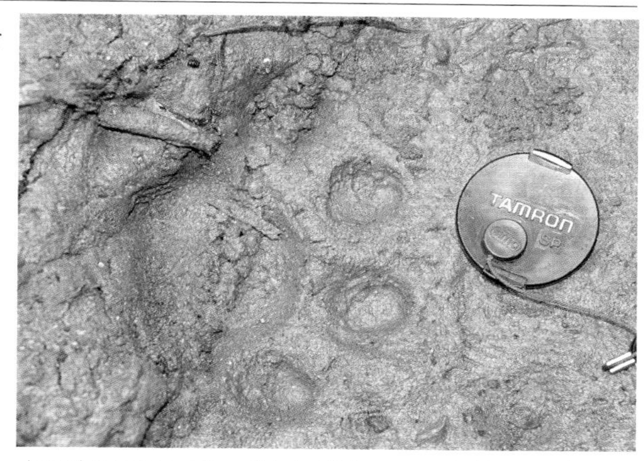

Jaguar forepaw track at Sirena Biological Station, compared with a camera lens cap two inches in diameter

Jaguar, emerging from cover

White-lipped Peccary adult

WHITE-LIPPED PECCARY

Tayassu pecari

Costa Rican name: *Cariblanco.*
2/23 trips; 2 sightings.

Total length: 45.3–49.2 inches, including 2.0-inch tail.

Weight: 55 pounds 2 ounces–75 pounds (25.0–34.0 kilograms).

Range: Southern Mexico to northern Argentina.

Elevational range: Sea level to 4,900 feet.

Like the tapir and jaguar, the endangered White-lipped Peccary is an indicator of wilderness rainforest habitat. Where large rainforest tracts exist, this wild pig occurs in roving herds of up to a couple hundred individuals. Early rainforest explorers feared encounters with this peccary because of its reputation for aggressive behavior, but there have been no recent reports of aggression by this species toward humans in Costa Rica. In spite of its ability to defend itself from other predators, this peccary is one of the first species to disappear when rainforest habitat is opened up with roads and then it is subjected to poaching.

The White-lipped Peccary roams the forest in search of palm nuts, fruits, roots, tubers, and invertebrates. It can eat very hard palm nuts by crushing them with its massive molars. The tusks are effective in defense against jaguars and pumas. When threatened, these peccaries make loud clacking noises with their tusks, and the long hairs of the back are erected to make the hogs look more intimidating. They also emit a powerful musky odor from scent glands on the back.

The peccary is active throughout the day. The approach of a herd can be ominous. There is a clattering noise from the tusks and the rustling of many feet through the forest litter. A strong, musky odor fills the air and remains long after the animals have passed. If approached by peccaries, one's best escape strategy is to climb a tree. Since peccaries can't jump, it is necessary to climb only about three feet to get above them.

Female peccaries mature at eighteen months of age, and they usually have two young 156 days after mating. The precocial young follow the mother soon after birth. The best remaining populations of White-lipped Peccaries are in Corcovado NP.

Skull of White-lipped Peccary adult, showing heavy molars used to crush seeds and canines used for defense

COLLARED PECCARY

Collared Peccary adult

As the name indicates, this peccary is marked by a faint whitish collar over the shoulders, which distinquishes it from the dark White-lipped Peccary. It is adapted to a much greater range of habitats and is found at higher elevations. Because of intensive hunting pressure, this mammal has been greatly reduced outside national parks and protected reserves.

The Collared Peccary is found in Guanacaste's dry forests, moist and wet lowland rainforests of the Caribbean and Pacific slopes, and montane forests. It can exist in disturbed rainforest from which White-lipped Peccaries have disappeared.

Reproductive details are similar to those of the White-lipped Peccary. Though the Collared Peccary is not often seen, the musk hangs heavy along forest trails where a herd has recently passed. Their presence is also indicated by disturbed soil where they have rooted for tubers, roots, snakes, and insect larvae. Many mammals in tropical forests assist in dispersing seeds that pass undamaged through their digestive tracts, but the peccary is an exception. It chews seeds with its molars before swallowing them, so they are seed destroyers. This is a survival problem for many plants, but not for the rainforest yam.

The rainforest yam (*Dioscorea*) contains diosgenin, which is a type of pig progesterone, or birth control for peccaries. If peccaries eat this yam, they stop having young, and the yam eaters die out. It is a natural form of selective peccary birth control that helps the yam survive in a forest full of peccaries. The yam extract was known by native Americans in Mexico and used for birth control, because pig progesterone is similar to human progesterone. It was reported that once a month they gave a drink with the yam extract to young unmarried girls in their villages to prevent them from having babies before they married. A Penn State University chemist, Russell Marker, checked out this rumor in 1940 and identified the source of this medicine. He formed a company named Syntex,

Tayassu tajacu

Costa Rican names: *Saíno; zahíno; chancho de monte; saíno de collar.*

5/23 trips; 9 sightings.

Total length: 34.3–37.0 inches, including 0.8–2.2-inch tail.

Weight: 30 pounds 14 ounces–55 pounds 2 ounces (14–25 kilograms).

Range: Arizona, New Mexico, and Texas to northern Argentina.

Elevational range: Sea level to 9,800 feet.

Rainforest yam, which contains a form of progesterone that prevents peccaries from reproducing

and in 1951 one of their chemists, Carl Djerassi, developed the first human birth control pills from the rainforest yam.

The Collared Peccary may occur in herds of up to thirty, but a typical herd is from two to fifteen. It is not as aggressive as the White-lipped Peccary, but observers should keep their distance. The best place to see this species is along the trails of La Selva Biological Field Station, where protection from poaching has allowed it to become abundant. Researchers there appreciate the peccaries because they are predators of snakes, including venomous snakes like bushmasters. The unfortunate side of this recovery is that the peccaries destroy the eggs and young of the rare White-fronted Nunbird, which makes its nest in a leafy tunnel under the leaf litter of the forest floor.

WHITE-TAILED DEER

Odocoileus virginianus

Costa Rican names: *Venado; venado cola blanca.*

5/23 trips; 6 sightings.

Total length: 59.1–86.6 inches, including 3.9–9.8-inch tail.

Weight: 50–80 pounds (22.7–36.3 kilograms).

Range: Southern Canada to northern Brazil.

Elevational range: Sea level to at least 7,200 feet.

It may seem surprising to encounter White-tailed Deer in tropical Costa Rica. This deer originated in North America and dispersed through Mexico and Central America to South America. Like raccoons and pumas, the White-tailed Deer demonstrates Bergmann's rule. Deer near the equator have a smaller body than the same species in colder northern temperate climates. This adaptation helps preserve body heat for larger northern deer during cold weather, and the smaller body size helps in dissipating heat for smaller deer in the tropics.

The White-tailed Deer is found in lowland and middle-elevation habitats where land clearing and agriculture have improved edge habitats and where early successional stages of vegetation provide twigs, leaves, fallen fruits, and nuts as food. They are not a species of undisturbed rainforest.

Bucks are in rut in Guanacaste from approximately July through November, and one or two fawns are born after a gestation period of 205 to 215 days. The antlers of bucks are usually no more than ten inches long along main beams that curve upward in an arching pattern. There are usually no more than two or three short tines along the beams.

This deer has been hunted so intensively in most regions, often at night with the aid of dogs, that it is rarely seen except in protected dry forest habitats of Guanacaste. It is common in Palo Verde, Santa Rosa, and Guanacaste NPs, and some may be observed in the vicinity of La Pacífica Biological Field Station.

White-tailed Deer adult doe

Red Brocket Deer adult buck. Photographed in Venezuela

RED BROCKET DEER

The elusive Red Brocket Deer is a small forest-dwelling species found in dense rainforest environments from Costa Rica's lowlands to the mountains. Its Spanish name, *Cabro de monte* ("goat of the mountain"), refers to the two short, straight goatlike antlers that are characteristic of bucks. This deer stands only about thirty inches high at the shoulders, and its back has a hunched posture that looks more like an agouti than a deer. The head is held low as it sneaks through heavy forest vegetation. In contrast to the more brownish-gray coat of a White-tailed Deer, a Red Brocket Deer has a deeper Hereford red coat with dark to blackish legs below the knees.

Mazama americana

Costa Rican name: *Cabro de monte.*

2/23 trips; 2 sightings.

Length: 39.1–53.3 inches, including 3.7–5.7-inch tail.

Weight: 26 pounds 6.4 ounces–70 pounds 6.4 ounces (12–32 kilograms).

Range: Tamaulipas, Mexico, to northern Argentina.

Elevational range: Sea level to 8,400 feet.

The Red Brocket Deer is usually solitary and may be active day or night, but it is most likely to be seen at dusk or at night with the aid of lights. Heavy hunting pressure in the past with the use of dogs has made this a shy species in settled areas. Even in unhunted habitats like national parks, a brocket deer may stand quietly as people pass by and remain unseen, or it may sneak away undetected.

The diet of brocket deer includes fruits, flowers, tender shoots, twigs, and fungi. Brocket deer have been observed in Tortuguero NP along canals during periods of high water when they were forced onto higher land. The deer shown here was photographed by Dr. Daniel H. Janzen in April 1965, about ten miles north of Palmar Sur. The other photo shows a Red Brocket Deer from the llanos of Venezuela, with the typical conformation and antler structure of the species, but the coat is paler than that of the forest-dwelling brocket deer of Costa Rica.

Red Brocket Deer adult buck. Photographed north of Palmar Sur, 1965. Photo © Dan Janzen

BAIRD'S TAPIR

Baird's Tapir adult female, with young, near Sirena Biological Station

The endangered Baird's Tapir is the largest mammal in Costa Rica and the largest of three tapirs in the Americas. Although it resembles a large pig, the tapir's closest relatives are horses and rhinoceroses. The tapir has well-developed incisor teeth like a horse and can bite viciously if attacked. Tapirs evolved in Asia and North America and spread southward into Central and northern South America after the creation of the Central American land bridge between North and South America three to four million years ago.

Adult tapirs are brown to blackish, with white tips on their ears. The tapir is incredibly adaptable and is found in habitats ranging from dry and riparian forests in Guanacaste to rainforests in the southern Pacific and Caribbean lowlands, cloud forests of Monteverde, and montane forests in the Talamanca Mountains. A creature of habit, it uses the same trails, defecation sites, stream-crossing sites, mud wallows, and feeding sites day after day. This predictability makes it vulnerable to poaching.

Active in the daytime and at night, the tapir is usually found near water and is an excellent swimmer. The diet consists of leaves, twigs, fruits, and seeds. The highly flexible snout aids in grasping and breaking off leaves and twigs while feeding. Since some seeds pass through the digestive tract undamaged, the tapir is an important seed disperser for tropical forest plants.

A single young tapir, camouflaged by patterns of white lines and white spots, is born after a gestation period of 390–400 days. The young stay with the female for up to a year, and the female bears young once every seventeen months.

Although Baird's Tapir is found throughout much of Costa Rica's national park system, it is rarely seen. The best place to look for it is at mud wallows, tide pools, and riverbanks in the vicinity of Sirena Biological Station in Corcovado NP on the Osa Peninsula.

Tapirus bairdii

Costa Rican names: *Danta*; *danto*.

1/23 trips; 1 sighting.

Total length: 79.5 inches, with a 2.8-inch tail.

Weight: 330–660 pounds (150–300 kilograms).

Range: Veracruz, Mexico, to the western Andes in Ecuador.

Elevational range: Sea level to 11,500 feet.

Plaster cast of a tapir's front foot track, showing three toe prints

Baird's Tapir

BALEEN WHALE FAMILY *(Balaenopteridae)*

HUMPBACK WHALE

Megaptera novaeangliae

Costa Rican name: *Ballena.*

1/15 trips; 1 sighting.

Total length: 36–50 feet.

Weight: 20–44 tons (18,200–40,000 kilograms).

Range: All oceans of the world.

The magnificent Humpback Whale is a migratory marine mammal that feeds in polar regions in summer and returns to warmer waters in winter. The pectoral fins are long and distinctive, and the dorsal fin is relatively small. White marks on the underside of the tail allow identification of individual whales. The back of the whale is black, and the throat and chest are white. Humpback Whales live in pods of up to a dozen individuals. Migratory groups may include up to 150 whales.

In polar areas, humpbacks feed on krill and other planktonic varieties of crustaceans. They also eat cod, capelin, and herring. The humpback is well known for the highly developed vocalizations by which it communicates with others of its own species. A single young is born after a gestation period of eleven months, and the interval between births is usually two years. This whale reaches sexual maturity at the age of ten years.

Until recently, Costa Rica was not known for whale watching. Humpbacks, however, can be seen in the Pacific Ocean between the Drake Bay Wilderness Resort and the Marenco Beach and Rainforest Lodge on the Osa

Humpback Whale surfacing near Caño Island

Peninsula and near the Caño Island BR during January and February. Those whales are migrants from the California population, and the females apparently come to Costa Rica for calving. Whales also winter in the Golfo Dulce area east of the Osa Peninsula. Watch for them while flying to or from Tiskita Jungle Lodge on the eastern shore of the Golfo Dulce. Other whale habitats include the Uvita Marine National Park south of Dominical and the Ballena Marine National Park on the Pacific coast, where whales winter from December through April.

The Ballena Marine National Park was created in 1991 by president Oscar Arias after two photos of whale tails taken by the author on a Henderson Birding Tour in 1990 were matched with photos taken offshore from California (archived by the Cousteau Society). This provided the documention needed to prove that Costa Rica was a migratory site for winter calving grounds of humpbacks offshore from Caño Island and Punta Uvita. The national park includes 425 acres of oceanfront land and 12,750 acres of ocean. For more information about whales, whale and dolphin conservation, and whale watching in Costa Rica, contact the conservation group Promar at 506-2233-9074 or pem@promar.or.cr, or visit www.fundacionpromar.org.

Humpback Whale diving near the Marenco Rainforest Lodge

Humpback Whales along the Pacific coast of Costa Rica, in 1990. By establishing the presence of humpbacks in the area, this photo resulted in the creation of Ballena Marine National Park by Costa Rican president Oscar Arias.

Red-eyed Tree Frog

AMPHIBIANS

There are 190 amphibians known in Costa Rica, representing members of all three orders in the class Amphibia: Gymnophiona (caecilians), Caudata (salamanders), and Anura (frogs and toads). Caecilians are small, wormlike vertebrates that have rudimentary eyes, a skull, and teeth. They inhabit leaf litter, burrow in the soil, and also live under fallen logs. Eight species of caecilians are found in Costa Rica. About forty-four species of salamanders are known from Costa Rica. Like caecilians, they are mainly fossorial, meaning that they live by burrowing, usually under a layer of fallen leaves, detritus, or fallen logs on the forest floor. Some may be found concealed in trees or low shrubs. Caecilians and salamanders are rarely seen or encountered by tourists.

Frogs and toads are much more likely to be seen by visitors. There are 140 species of frogs and toads in Costa Rica, and 22 of them are treated in the following accounts. Some were selected because they are more conspicuous, and some are included for their stunning appearance and interesting natural history.

Amphibians have incredible adaptations for reproduction and survival. The reproductive behavior of certain frogs enables them to avoid raising tadpoles in ponds inhabited by fish. Because fish eat frog eggs and tadpoles, survival chances are increased for frogs that can find fishless ponds in which to raise their young. For instance, poison dart frogs lay their eggs in forest litter and then transport them to the water tanks of bromeliads or other small bodies of water, and some frogs lay their eggs in moist ground litter where their tadpoles can hatch and grow without being eaten by fish. Direct development is another extreme adaptation that makes frogs such an amazing component of Costa Rica's biodiversity. "Direct development" means that a frog embryo develops completely within the egg, without hatching into a larval or tadpole stage. Such a frog hatches as a froglet.

Some amphibians are camouflaged with cryptic markings for survival. Others, like dendrobatid poison frogs, are brightly colored to warn predators of their toxicity. Survival strategies include the semitransparency of glass frogs, which allows them to blend with their background, and the toxic chemicals that many toads have in their skin to discourage predation.

Don't rush to judgment on the identity of most amphibians. Individual species can vary greatly in their appearance and colors from one region or watershed to the next, and, within a single genus there can be a dozen or more similar species. Some frogs change color during the day, depending on whether they are active or resting. In most cases, you must be content to identify the genus of an amphibian. You can enjoy their beauty, behavior, and adaptations regardless of their identity.

Avoid touching or grabbing frogs; you may have sunscreen, insect repellent, or other chemicals on your hands that could injure or possibly kill them. Some toads and frogs can also cause extreme reactions on your skin.

The endemic Golden Toads of Monteverde are not included among these accounts because they have not been observed since 1988 and are believed extinct. Current evidence points to the spread of a deadly chytrid fungus through Costa Rica in 1986 and 1987 as the cause of the extinction.

Of more immediate concern is global warming and the warming and drying effect on high-humidity cloud forest habitats like the Monteverde rainforest. Dry forest wildlife species are becoming more abundant at higher elevations, and amphibians are among the first species to succumb to dramatic changes in habitat. These changes are expected to continue. Amphibians have become among the most critical indicators of water-quality problems and climate change not only in Costa Rica but also in many other regions around the world.

While the Monteverde area is becoming drier and amphibian populations are either declining or at risk, one area in the lower Caribbean foothills has emerged as a new mecca for amphibian diversity and research in the country, the Costa Rican Amphibian Research Center. The site has the highest diversity of amphibians in Costa Rica. Located in the eastern Caribbean foothills of the Talamanca Mountains near the town of Guayacán, south of Siquirres, the center is a working research station, where more than fifty amphibian species have been documented, including many very rare species. If you are interested in tropical amphibians, the research center is a must-see destination. It is owned and operated by herpetologist Brian Kubicki. If you would like to learn more about amphibians in Costa Rica, check out his Web site at www.cramphibian.com.

Sincere appreciation is extended to Brian Kubicki for reviewing and editing the following accounts of the frogs and toads and graciously allowing the use of his outstanding photos of many of those frogs and toads.

LITTER TOAD

Rhaebo haematiticus (formerly Bufo haematiticus)

Costa Rican name: *Sapo.*

Total length: 2.0–2.8 inches (51–70 millimeters).

Range: Eastern Honduras to central Ecuador.

Elevational range: Sea level to 4,200 feet.

A small smooth-skinned toad, *Rhaebo haematiticus* has no cranial crests and is coppery brown above with a thin copper-colored line that passes from the nose, above each eye, and along each side of the body to the base of the hind legs. There are one to four small, dark brown, paired crescent or apostrophe-shaped marks on the back. Huge parotoid glands on the sides of the neck produce milky venom that deters predators. When calling, the male produces a series of chirps that sound like a chick in search of food.

This toad is found in moist and wet lowland and middle-elevation forests of the Caribbean and southern Pacific regions, including La Selva Biological Field Station, the Costa Rican Amphibian Research Center, Drake Bay Wilderness Resort, Corcovado NP, and Wilson Botanical Garden. Strings of eggs are laid in small pools of water that remain in stream beds during the dry season. Adults are found on the forest floor foraging for small insects and other arthropods in the leaf litter.

Litter Toad. Photo © Brian Kubicki

GIANT TOAD (MARINE TOAD)

Rhinella marina (formerly Bufo marinus)

Costa Rican name: *Sapo grande.*

Total length: Up to 7.9 inches (20.1 cm).

Weight: 2.6 pounds.

Range: Southern Texas to central Brazil.

Elevational range: Sea level to 6,600 feet.

This huge, well-known toad is the largest toad in the Americas. It is most common in yards, gardens, farms, and human settlements. The very large parotoid glands, one on each side of the neck, contain a toxic chemical that deters predators; it is powerful enough to kill a dog. This toad has another effective defense: if you pick one up, it copiously urinates on you. This amphibian has become an exotic pest in many tropical and subtropical countries where it has been introduced by humans.

Rhinella marina is best known for its enormous appetite. It seems to eat just about anything: small mammals, smaller Giant Toads, spiders, hemipteran bugs, beetles, millipedes, ants, wasps, and vegetable matter. Natural population densities may be about 10 per acre, but populations around human dwellings exceed 100 per acre. Giant Toads remain concealed during the day under logs, rocks, and other objects and emerge at night to feed, often under lights. Individual toads feed only once every three to four days. They are most active for the first several hours after dark.

This toad is very prolific. At the end of the dry season, it lays masses of eggs in shallow edges of pools and ponds. The number of eggs laid may vary from 5,000 to 25,000, depending on the size of the female. A toxic coating on the eggs protects them from predation. This toad reaches sexual maturity at one year of age. Females grow rapidly for three to four years and become larger than males, whose growth slows after one year. As with all amphibians, these toads continue to grow throughout their lives.

Giant Toad. Photo © Brian Kubicki

WET FOREST TOAD

Wet Forest Toad. Photo © Brian Kubicki

The Wet Forest Toad is characterized by a prominent cranial crest and a dark broad stripe on the side; there is a border of light-colored warts above the stripe. The parotoid gland is small and triangular.

This toad's preferred habitat includes moist and wet rainforests and premontane forests of the Caribbean lowlands. Look for this nocturnal toad in forest litter or along stream edges. It breeds during the drier season from January through March in pools along rocky streams, and the species is more closely associated with streams at that time.

This toad eats small invertebrates, but its biology is not well known. The toad shown below was photographed along the forest loop trail at Rancho Naturalista. It can also be seen at the Costa Rican Amphibian Research Center.

Incilius melanochlorus (formerly Bufo melanochlorus)

Total length: 1.7–4.2 inches (43–107 millimeters).

Range: Regional endemic on the Caribbean slopes and lowlands of Costa Rica into northwestern Panama.

Elevational range: Sea level to 3,240 feet.

Wet Forest Toad

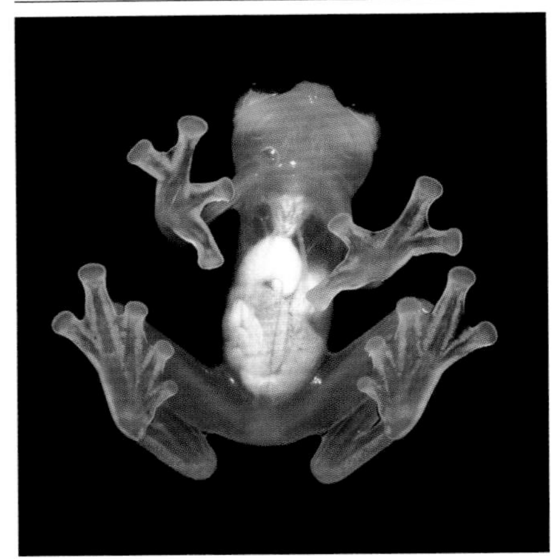

Glass frog *Hyalinobatrachium valerioi*, showing transparent body and internal organs. Photo © Brian Kubicki

Genera: Cochranella, Espadarana, Hyalinobatrachium, Sachatamia, and *Teratohyla*

Costa Rican name: *Ranita de vidrio.*

Total length: Up to 1.5 inches (38 millimeters).

Range: Southern Mexico to Paraguay and northern Argentina.

Elevational range: Sea level to over 6,000 feet.

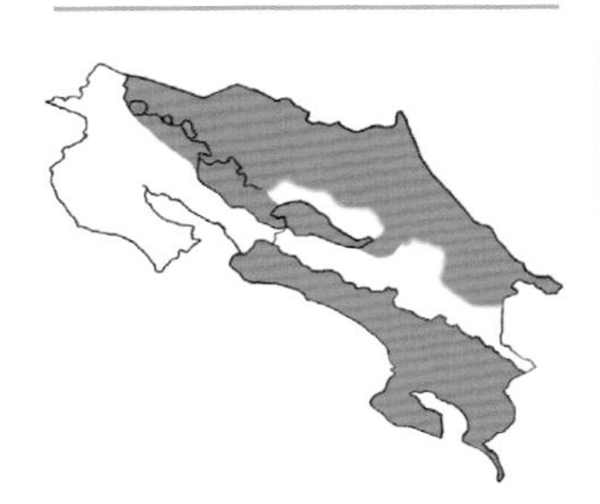

GLASS FROGS

In Costa Rica five genera and thirteen species of glass frogs are currently recognized to inhabit the humid forest regions. They include *Cocharanella* (*C. euknemos* and *C. granulosa*), *Espadarana* (*E. prosoblepon*), *Hyalinobatrachium* (*H. chirripoi*, *H. colymbiphyllum*, *H. fleischmanni*, *H. talamancae*, *H. valerioi*, and *H. vireovittatum*), *Sachatamia* (*S. albomaculata* and *S. ilex*), and *Teratohyla* (*T. pulverata*, and *T. spinosa*). Two of these species, *H. chirripoi* and *H. talamancae*, have been recently rediscovered, nearly fifty years after their last collection in Costa Rica.

In contrast to poison dart frogs that advertise their presence with bright colors, glass frogs are protected from predation by a greenish translucence that allows their body to blend with the color of their leafy background. Through the transparent skin on some species can be seen the beating heart as well as the stomach, liver, gall bladder, ventral vein, intestines, and lungs.

The majority of Costa Rica's glass frogs are found along streams in the humid forests of the lowlands and foothills of the Caribbean and Pacific slopes. Two species, *Espadarana prosoblepon* and *Hyalinobatrachium fleischmanni*, can also be heard and less often seen in the streams of the Central Valley and other higher-elevation humid forest regions of Costa Rica. These two species have the largest altitudinal range in Costa Rica, being found from sea level

Glass frog *Teratohyla spinosa*

to above 6,000 feet, which is also the upper limit for glass frogs in Costa Rica.

Glass frogs are active on vegetation along streams and rivers during the night. The males call from territories along the streams and rivers, and depending on the species, they may be observed calling from the top of a leaf or from the underside. Some species occasionally use rocks, branches, or logs as calling sites. The males call to define their territory and in hopes of attracting a mate. Once a female approaches a male, mating (technically referred to as axillary amplexus) follows, and eventually the pair will produce a delicate gelatinous egg mass overhanging the water. Depending on the species, each mass may contain twenty to eighty or more eggs. The eggs are normally deposited on the upper or underside of a leaf, but at times egg masses can be found on rocks, on walls near waterfalls, or even on mossy branches. The form, number of eggs, and position of the mass varies from one species to another, and in most cases it is possible to identify the species that produced an egg mass based on those characteristics.

Glass frog *Hyalinobatrachium valerioi* male, guarding an egg mass. Photo © Brian Kubicki

Tadpoles hatch from the egg mass after about fifteen days, and then they fall into the stream or river below. The tadpoles undergo their larval development in the bottom of the stream. It may take a year or more before one emerges from the water as a little green froglet with a long muscular tail. The tail is slowly resorbed into the body.

Glass frog *Sachatamia ilex*. Photo © Brian Kubicki

Some species of glass frogs give their egg masses specialized parental care. In Costa Rica, such care has been documented in *H. valerioi*, *H. talamancae*, *H. chirripoi*, *H. vireovittatum*, *H. colymbiphyllum*, and *H. fleischmanni*. The males of these frogs guard the egg masses during the evening. In some of these species, such as *H. valerioi*, *H. talamancae*, and *H. chirripoi*, the males will also guard the egg masses during the day.

Glass frogs can be hard to find, but one of the best places to look for them is in the vegetation along small forested streams. The best conditions are in the evening following a rain shower. A good place to see glass frogs is the Costa Rican Amphibian Research Center, where eight species are known to inhabit the streams of this private reserve.

For more information about the remarkable family of glass frogs, the reader is referred to the book *Ranas de Vidrio (Glass Frogs) of Costa Rica*, by Brian Kubicki.

RAIN FROG (FITZINGER'S LITTER FROG)

Craugastor fitzingeri (formerly Eleutherodactylus fitzingeri)

Costa Rican name: *Rana de lluvia; ranita piedrita.*

Total length: 2.0 inches (51 millimeters).

Range: Northeastern Honduras to central Colombia.

Elevational range: Sea level to 5,000 feet.

The Rain Frog is a common and widely distributed frog found the length of the Pacific and Caribbean lowlands. The body is a light to medium sandy brown, with a fine pebbly texture. Many show narrow black bands that begin at the point of the nose and extend through the eyes and the top of the tympanum. This band then turns downward and ends just behind the tympanum. Some individuals have a narrow light stripe down the center of the back, and most show indistinct, parallel brownish bands across the thighs. The Rain Frog also shows a prominent gold horizontal bar through the top edge of the eye. The best distinguishing mark for this highly variable frog is the yellow spots on the back of the thighs.

This terrestrial frog can be encountered in Tortuguero NP, including the trails at Tortuga Lodge, and also in Cahuita NP. It occurs in Corcovado NP and can be discovered in leaf litter along the trails at Sirena Biological Station. Other locations include La Selva Biological Field Station, Carara and Manuel Antonio NPs, the Rainforest Aerial Tram property, lower levels of Braulio Carrillo NP, and the Costa Rican Amphibian Research Center. During the day, it can be found in leaf litter, and at night it reveals its presence by calling as it rests on leaves of low vegetation.

Rain Frog. Photo © Brian Kubicki

BRANSFORD'S LITTER FROG

Craugastor bransfordii (formerly Eleutherodactylus bransfordii)
Costa Rican name: *Rana (sapito de hojarasca).*
Total length: 1.2 inches (31 millimeters).
Range: Nicaragua to Panama.
Elevational range: Sea level to 3,900 feet.

This frog, perhaps the most common frog in Costa Rica, is found on the forest floor throughout lowlands and middle elevations in moist and wet forests of the Caribbean slope of Costa Rica. Many colors and skin textures are known for this frog, which provide camouflage on widely varying colors and textures of forest litter. The posterior edge of the thigh is reddish.

This diurnal frog eats ants, mites, beetles, hemipteran bugs, insect larvae, and spiders. *Craugastor bransfordii* breeds at night and lays eggs on moist sites on the ground where there is enough moisture to keep them from drying out. The embryos develop directly within the eggs and hatch as froglets. This frog can be seen at La Selva Biological Field Station and the Costa Rican Amphibian Research Center. In the southwest Pacific region, the similar *C. stejnegerianus* can be seen at the Wilson Botanical Garden and the Sirena Biological Station in Corcovado NP.

Bransford's Litter Frog. Photo © Brian Kubicki

ROBBER FROG

Craugastor podiciferus (formerly Eleutherodactylus podiciferus)

Total length: 0.8–1.6 inches (21–40 millimeters).

Range: Costa Rica and western Panama.

Elevational range: 3,250–7,950 feet.

This well-camouflaged diurnal frog is found in the leaf litter of mature forests. Its life history is poorly known, but this genus of frogs eats small invertebrates like mites, spiders, centipedes, isopods, ants, katydids, and beetles. This is a regionally endemic species, found only in the mountains of Costa Rica and adjacent mountains of western Panama.

Craugastor podiciferus is tan to dark brown, with a black mask from the side of the nose through the eyes, and it has slender fingers. It is a high-elevation species, found from premontane to lower montane levels of Monteverde, volcanoes surrounding the Central Plateau, and the Talamanca Mountains from the Central Plateau to Panama. It has been observed in forests of the San Gerardo de Dota Valley at approximately 7,600 feet.

Robber Frog

SMOKY JUNGLE FROG

Smoky Jungle Frog

The Smoky Jungle Frog is the largest frog in Costa Rica. Among amphibians, it is second only to the Giant Toad (*Rhinella marina*). It is distinguished by its huge size, medium brown back and sides, brown-to-black splotches along the upper lip, narrow black stripe through the eye that extends above the tympanum, yellowish belly, and symmetrical golden brown squiggly lines on the back and sides. It has a large gland in the groin area. Frequently found near streams and ponds, this frog inhabits burrows in the ground and in rock crevices. Habitats include dry and riparian forests in Guanacaste and moist and wet lowlands and middle elevations of both the Caribbean and Pacific regions.

The species epithet of this frog was recently changed to honor pioneer Costa Rican herpetologist Dr. Jay M. Savage, the author of the definitive book *The Amphibians and Reptiles of Costa Rica*, published in 2002.

This frog has an appetite for other frogs, insects, crabs, crayfish, and snakes up to twenty inches long. It has interesting defenses. The skin secretes very irritating toxins that cause a predator to drop it quickly. It gives out a loud scream that may serve to warn other frogs, but the noise also seems to attract nearby caimans. It is believed that the approach of a caiman may subsequently scare off the frog's predator.

Eggs are laid in huge foam masses that can exceed a gallon of foam. The masses are located on the edge of temporary woodland pools or dry depressions in stream beds where they are not exposed to predatory fish. The foam prevents the eggs from drying out while the tadpoles develop. The partially carnivorous tadpoles eat the eggs and tadpoles of other frogs. This frog may be encountered in Corcovado NP, Lapa Ríos, Corcovado Lodge Tent Camp, Tiskita Jungle Lodge, and the Costa Rican Amphibian Research Center.

Leptodactylus savagei (formerly Leptodactylus pentadactylus)

Costa Rican name: *Rana ternero.*

Total length: Up to 6.3 inches (160 millimeters).

Range: Caribbean slope from northern Honduras to Colombia and Ecuador; Pacific slope from northern Nicaragua to Brazil.

Elevational range: Sea level to at least 3,900 feet.

Green and Black Poison Dart Frog.
Photo © Brian Kubicki

GREEN AND BLACK POISON DART FROG

Dendrobates auratus

Costa Rican name: *Rana venenosa.*

Total length: Up to 1.6 inches (41 millimeters).

Range: Southeastern Nicaragua to northwestern Colombia.

Elevational range: Sea level to 1,900 feet.

Poison dart frogs derive their name from the Golden Poison Dart Frog (*Phyllobates terribilis*) of western Colombia. Their skin contains an alkaloid (batrachotoxin) so dangerous that a two-inch frog has enough poison to kill eight humans. Chocó Indians of western Colombia treat their blowgun darts with juice from the skin of these frogs. Although other frogs in this family are not used to treat blowgun darts, their skin contains a potent but milder venom called pumiliotoxin-C.

The Green and Black Poison Dart Frog inhabits humid forests along the lowlands and foothills of the Caribbean slope and the central and southern Pacific region. It is active throughout the day among leaf litter on the forest floor and on logs, stumps, trees, and shrubs to a height of fifty feet. A major portion of its diet consists of ants and termites.

Reproduction is a complex process by which the male calls a female to him on the forest floor. After laying four to six eggs, she leaves. The male covers the eggs with sperm and tends the fertilized eggs for the next four to thirteen days. Meanwhile, he calls more females, entices them to lay eggs, and fertilizes and takes care of those egg masses as well. During this period, he sits in small puddles to absorb water and then sits on the eggs to keep them moist. Females lay eggs about every two weeks throughout the breeding season.

As each egg hatches, the tadpole climbs up a hind leg of the male and onto its back. The tadpoles are transported one at a time to tree cavities, bromeliad water tanks, and water-filled cavities in logs. The tadpoles eat aquatic organisms there such as protozoans, rotifers, and mosquito larvae. They develop into frogs in about twelve weeks. This frog may live four years or more. It may be observed at the Costa Rican Amphibian Research Center, Corcovado NP, Lapa Ríos, and Sirena Biological Station, and on the grounds of Corcovado Lodge Tent Camp and Tiskita Jungle Lodge. It is best observed when ground conditions are moist to wet.

STRAWBERRY POISON DART FROG

Strawberry Poison Dart Frog

Oophaga pumilio (formerly Dendrobates pumilio)

Costa Rican name: *Ranita roja; rana venenosa.*

Total length: 0.8–0.9 inch (20–23 millimeters).

Range: Nicaragua to northwestern Panama.

Elevational range: Sea level to 3,200 feet.

The brightly colored Strawberry Poison Dart Frog demonstrates warning coloration to potential predators. If a snake bites such a frog, it immediately releases the frog, scrapes its mouth against the ground, and may writhe or lie comatose for several hours. Snakes, birds, and mammals do not die from this experience, but they learn not to eat brightly colored little frogs.

This frog has many color patterns, but the most famous is a bright red body with blue legs, sometimes referred to as the "red frog wearing blue jeans," the variation found in the Caribbean lowlands north of the Reventazón River. There is a continuing shift of different color patterns toward the Panama border.

This species is found in Caribbean lowland rainforests and is most active in the morning. Ants, termites, mites, and tiny insects make up its diet. It is believed that toxic pumiliotoxin-C in the skin may be derived from chemicals in the ants in its diet, because the toxicity of the skin significantly declines in captivity when the frog doesn't have access to rainforest ants.

The reproductive behavior of this frog is one of the most incredible stories in the rainforest. Males establish territories on logs and stumps at a spacing of about ten feet. Their mating call is a cricketlike buzz that pulses at a rate of four to five buzzes per second, deterring males while attracting females. If another male approaches, the two males rise up and grapple with each other like little sumo wrestlers. When a female approaches, the male leads her to a nesting site in the ground litter, where he deposits sperm on a leaf and she deposits two to five eggs on the sperm. He guards the eggs and keeps them moist for about seven days until they hatch.

When the eggs hatch, the female returns, and the tadpoles climb onto her back, clinging to her by using their mouths as suckers. She climbs trees and backs into the water tanks of bromeliads or water-filled plant cavities. The tadpoles slide into the water, and the female returns for the other tadpoles until she has placed them all in the water of other cavities. She visits each tadpole every one to nine days for the fifty days it takes to develop. When the tadpole senses its approaching mother, it vibrates its tail. Then she backs into the water and lays an unfertilized egg for the tadpole to eat. She provides seven to eleven eggs for each tadpole during its development. This process is an adaptation that tropical frogs have developed to avoid raising tadpoles in ponds that contain fish. This colorful little frog is an opportunist, however. It will also deposit tadpoles directly in the stagnant water of old tires, cans, and even water-filled beer bottles.

The most dependable places to view this frog are along the nature trail at Tortuga Lodge near Tortuguero, at the Costa Rican Amphibian Research Center, and at Sueño Azul. It may also be seen along the trails at La Selva Biological Field Station when rainy weather keeps the ground cover moist.

Two Strawberry Poison Dart Frogs meet in a territorial encounter.

GRANULAR POISON DART FROG

Oophaga granulifera (formerly Dendrobates granuliferus)

Total length: 0.8–0.9 inch (19–22 millimeters).

Range: Regional endemic in southwestern Costa Rica and western Panama.

Elevational range: Sea level to 2,300 feet.

The Granular Poison Dart Frog is endemic to Costa Rica and extreme western Panama. It is found in Pacific lowland rainforests from near Carara NP to the Golfo Dulce region. This frog, closely related to *Oophaga pumilio*, demonstrates a wide variety of color variations and patterns, including yellows, greens, and reds. Some are red or orange with green legs, sides, and belly. In some portions of the range, individuals have a green to yellow back with turquoise legs. Some are all olive-yellow, and others are a solid copper color. The skin has a granular texture, hence its name. The skin glands contain the toxin that is so potent in protecting the frog. The life history is similar to the Strawberry Poison Dart Frog. Look for this frog in Corcovado NP on the Osa Peninsula and along trails at Tiskita Jungle Lodge.

Granular Poison Dart Frog, red body with green legs

Granular Poison Dart Frog, lime green with blue toes. Photo © Brian Kubicki

Granular Poison Dart Frog, orange body with turquoise legs. Photo © Brian Kubicki

Granular Poison Dart Frog, green body with blue legs. Photo © Brian Kubicki

PHYLLOMEDUSINE SUBFAMILY *(Phyllomedusinae)*

RED-EYED TREE FROG (GAUDY TREE FROG)

Agalychnis callidryas

Costa Rican name: *Rana calzonuda.*

Total length: Up to 2.8 inches (71 millimeters).

Range: Southern Mexico to northwestern Colombia.

Elevational range: Sea level to 3,280 feet.

The Red-eyed Tree Frog is often featured in rainforest literature and in Costa Rican tourism brochures. The expressive, bright red eyes with black vertically elliptical pupils; green body; and blue, white, and orange highlights make it one of the most colorful rainforest creatures. This frog is fairly common in many of the humid forest regions of the lowlands and foothills of the Caribbean and central and southern Pacific regions. Unfortunately, it is rarely seen by tourists.

Adults inhabit the canopy of the forest during most of the year, but they come down to lower canopy levels with the onset of the rainy season in March to April. The males locate small, temporary rainforest pools (pools that lack predatory fish), and then they give a mating call that attracts the females. After the male mounts the female, the mating pair descends to the pool, where the female absorbs water through the skin into her bladder. (The male clings to the back of the female throughout the mating process, called axillary amplexus.) Then the pair climbs to the upper or underside of a leaf overhanging the

Red-eyed Tree Frog

pool, where the female lays a mass of 11 to 104 eggs during amplexus as the male releases sperm. The female releases the water from her bladder onto the newly fertilized eggs, creating a gelatinous foam that covers the eggs and keeps them from drying out. The female returns to the pool with the male still on her back, refills her bladder, and climbs back up to repeat the process. She may lay three to five clutches of eggs in one night. After five to eight days, the eggs hatch and the tadpoles drop from the gelatinous mass into the pond. They feed on suspended organic matter and develop into frogs in about eighty days.

Red-eyed Tree Frogs eat small insects, moths, and other arthropods. They are eaten by birds, coatis, kinkajous, and Fringe-lipped Bats (*Trachops cirrhosus*). The bats locate the frogs by the calls of the males. The adults and egg masses are also eaten by Cat-eyed Snakes (*Leptodeira septentrionalis*). Red-eyed Tree Frogs are very common at several of the ponds at the Costa Rican Amphibian Research Center. They can also sometimes be located at night on foliage adjacent to ponds in the courtyard at Tortuga Lodge and at Celeste Mountain Lodge. Another place to look for them (during guided night walks) is on emergent vegetation next to boardwalks over the woodland pools along Sendero Cantarana at La Selva Biological Field Station.

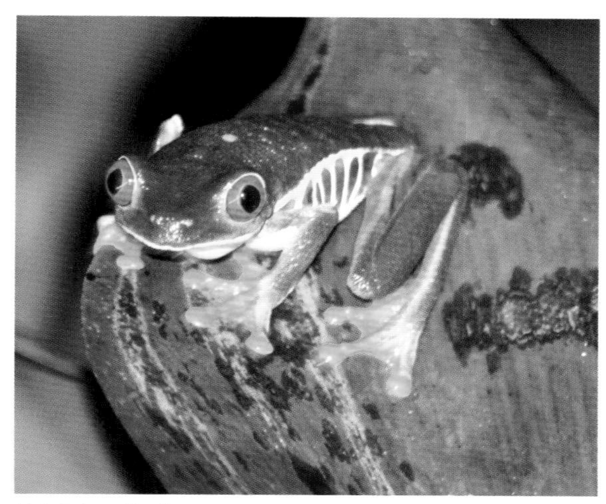

Adult, Red-eyed Tree Frog

GLIDING LEAF FROG

Adult, Gliding Leaf Frog

Agalychnis spurrelli

Total length: 2.0–3.7 inches (50–95 millimeters).

Range: Costa Rica and Panama to the Pacific lowlands of Colombia and Ecuador.

Elevational range: Sea level to 2,900 feet.

This frog's common name refers to its ability to leap from an aerial perch and glide downward with its body flattened and its legs and feet outspread, which would provide a ready escape route from a predator like a snake. There are two forms, possibly distinct species, in Costa Rica: the smaller form is found in the lowlands and foothills of the central and southern Pacific slope, and a recently discovered large form is from the lowlands and foothills of the central and southern Caribbean slope.

These frogs have a dark green back with solid yellow flanks. The Pacific form has a white ventral surface, and the larger form, from the Caribbean, has a yellow ventral surface. The iris in both forms is dark red to maroon. Numerous white spots are normally present on the back. The white dorsal spots have a fine black outline. Extensive interdigital webbing is present on the hands and feet, especially in specimens from the Caribbean slope.

The Gliding Leaf Frog inhabits old-growth and secondary humid forests. This uncommon species is currently known to exist at only a few sites in Costa Rica. To add to the difficulty in observing this species, it is an explosive breeder, normally found only near breeding sites for a few days at a time. It reproduces by laying its eggs on the vegetation overhanging ponds in secondary and old-growth forest. Brian Kubicki has documented that reproduction in this species is most often associated with the full moon. During breeding events, hundreds of adult individuals can be seen on the ground and vegetation surrounding a breeding pond. The peak breeding activity often takes place after midnight, when it is possible to see dozens of pairs in amplexus or ovipositing. At times in large breeding aggregations, masses of males can be seen wrestling as they try to enter amplexus with a single female. Embryonic development is normally completed after five days. Larval development can take up to six months. The tadpoles feed on suspended organic material in the water.

This rare species is difficult to observe in most areas, but a good place to see it is at the Costa Rican Amphibian Research Center during a full moon in the rainy season.

For more information about this species and other Costa Rican phyllomedusine frogs, the reader is referred to the book *Leaf Frogs of Costa Rica*, by Brian Kubicki.

SPLENDID LEAF FROG

Splendid Leaf Frog. Photo © Brian Kubicki

The Splendid Leaf Frog is without a doubt one of the most beautiful frogs in Costa Rica. This large tree frog has a dark green back with numerous light green spots. The flanks are yellowish orange with black vertical bars. Black stripes are also present on parts of the hands and feet. The eyes have a vertically elliptical pupil and a yellow and gray iris. On the tip of the heel is a fleshy structure known as the calcar, or spur, which is the source of the specific name *calcarifer*, Latin for "spur wearing."

In Costa Rica this uncommon nocturnal species is known only from the humid forests along the lowlands and foothills of the Caribbean slope. It is primarily found in old-growth forests, but at times it can be found in remnant patches of secondary forest. Males call during the evening from the vegetation near water-filled cavities in fallen or standing trees. The frog attaches its egg masses to vines, roots, leaves, or bark overhanging the water in these cavities.

Embryonic development normally lasts five to seven days, after which the tadpoles hatch, free themselves from the thick jelly of the egg mass, and fall into the water below.

Larval development takes from several months to over a year. It is believed that the young frogs then climb into the forest canopy to continue their development. It may take from a year to eighteen months to reach sexual maturity.

This is a difficult species to find because its reproductive habitat is so specific and so uncommon, and its soft call is easily missed. A good place to see the Splendid Leaf Frog is the Costa Rican Amphibian Research Center. For more information about this species and other Costa Rican phyllomedusine frogs, the reader is referred to the book *Leaf Frogs of Costa Rica*, by Brian Kubicki.

Cruziohyla calcarifer (formerly Agalychnis calcarifer)

Total length: 2.1–3.5 inches (55–90 millimeters).

Range: Honduras to Ecuador.

Elevational range: Sea level to 2,600 feet.

LEMUR LEAF FROG

Agalychnis lemur (formerly Phyllomedusa lemur, Hylomantis lemur)

Total length: 1.2–1.8 inches (30–45 millimeters).

Range: Scattered sites along the premontane Caribbean slopes of Costa Rica and Panama.

Elevational range: Historically 1,500–5,300 feet, but currently known in Costa Rica only at 1,500–1,750 feet.

The large eyes of the Lemur Leaf Frog make it one of the most charismatic frogs in Costa Rica. This small and delicate frog shows strong color changes between active periods and resting periods. When active, it has a dark red and green back and a dark gray iris. At rest, it normally has a light lime-green back with red spots and a light silver iris. The belly of the Lemur Leaf Frog is white. This species is distinguished from other Costa Rican phyllomedusine frogs by the lack of webbing between its toes.

The Lemur Leaf Frog was historically found at numerous mid-elevation humid forest sites along the Caribbean slope of Costa Rica, but in the late 1980s and early 1990s, the majority of the known populations disappeared. The species is now considered critically endangered in Costa Rica, being known at only two sites, both in Limón Province. The first site is in a patch of forest near Limón, and the second is the Costa Rican Amphibian Research Center.

This frog reproduces by laying its eggs on vegetation overhanging small pools or above the slow-moving water of seepages and springs. Breeding takes place throughout much of the year in the two existing sites, where rains continue throughout most of the year.

Because of its rarity, the only place where it is possible to see this frog in Costa Rica is at the Costa Rican Amphibian Research Center. For more information about this species and other Costa Rican phyllomedusine frogs, the reader is referred to the book *Leaf Frogs of Costa Rica*, by Brian Kubicki.

Lemur Leaf Frog, active coloration. Photo © Brian Kubicki

Lemur Leaf Frog, resting coloration. Photo © Brian Kubicki

HYLINANINE FROG SUBFAMILY *(Hylinae)*

MILK FROG

Trachycephalus venulosus (formerly Phrynohyas venulosa)

Total length: 2.8–4.5 inches (70–114 millimeters).

Range: Caribbean slope from Nicaragua to central Panama, but mainly occurs on the Pacific slope from northwestern Costa Rica to northwestern Ecuador.

Elevational range: Sea level to 2,100 feet.

This large, nocturnal hylid frog derives its name from the milky secretion that oozes from glands in its skin if it is handled or captured by a predator. The secretion is an alkaline compound that is irritating to mucous membranes. If you pick one up and then rub your eyes, it can be extremely painful and cause temporary blindness.

Among the distinctive characteristics of this frog is a pair of large, balloonlike vocal sacs at the sides of the male's jaws. The call is described as a repetitive series of loud growls. This frog is a parachuting specialist, in that it can jump from a tree branch, extend its legs, flatten its body, and sail to another location. In one instance, a Milk Frog jumped from a perch 130 feet high and landed 80 feet away, measured horizontally. This is an aid in escaping from predators and moving to new foraging habitats.

Milk Frogs inhabit bromeliads, heliconias, bananas, and tree holes. They emerge at night to feed on small invertebrates as they perch on leaves or branches. They are adapted to life in tropical dry, moist, and wet forests.

Look for this nocturnal species at night on low shrubbery and vegetation. The specimen shown here was found in the courtyard at Villa Lapas near Carara NP.

Milk Frog

MASKED TREE FROG

Smilisca phaeota
Total length: 1.6–3.0 inches (40–78 millimeters).
Range: Honduras to western Ecuador.
Elevational range: Sea level to 3,350 feet.

A prominent black marking that extends from the eye back to the shoulders gives this nocturnal tree frog its name. It is characterized by a green area between the eye and nostril and a white upper lip. The overall body color is light tan to mint green, but markings vary considerably. The call is described as a loud "wraaap-wrop."

This tree frog lives in the lower levels of moist and wet tropical forests. During the day it remains hidden in the curled up leaves of heliconias and bananas or on vegetation where it is concealed by an overhanging leaf. It eats small invertebrates. Habitat includes vegetation near swampy areas, small ponds, streams, ditches, and even backyards where water and drain tiles provide hiding places. The Masked Tree Frog breeds during the rainy season, laying more than 1,500 eggs in a film on the surface of ponds beside emergent plants in the water, on pools, and in tire tracks. The males call while floating on the water's surface.

This frog can be found in lowland and premontane levels of both the Caribbean and southern Pacific moist and wet forests. It is common at the Costa Rican Amphibian Research Center. Since it is nocturnal, it may be seen on the foliage of plantings at ecolodges by searching courtyards after dark with the aid of flashlights. One must be cautious, however, about other venomous creatures that are out at night.

Masked Tree Frog. Photo © Brian Kubicki

DRAB TREE FROG

This nocturnal tree frog spends most of the day sleeping and concealed in vegetation. It emerges at night on rocks or stones in streams, where it makes a call described as a repetitive series of "wrink" sounds that increase in inflection toward the end of the call. Individuals may be gray, tan, reddish brown, or golden brown. It has conspicuous webbing between its toes.

Prey is believed to consist of small invertebrates, and its preferred habitat includes river and streamside vegetation amid rocks, stones, and pebbles. During the breeding season males may be found at night calling from rocks in streams or along gravelly edges of streams. Calling and breeding activity is more pronounced after heavy rains. Eggs are laid in shallow forest rainwater pools and in shallow slow-moving or still sections of streams and rivers.

The distribution of the Drab Tree Frog includes Caribbean and southern Pacific lowland and premontane moist and wet forests, the Central Plateau, and Guanacaste forests east and northeast of the Gulf of Nicoya, but it is not found in the dry forests of Guanacaste. It can be seen at the Costa Rican Amphibian Research Center.

Smilisca sordida

Total length: 1.3–2.5 inches (32–64 millimeters).

Range: Honduras to western Panama.

Elevational range: Sea level to 4,575 feet.

Drab Tree Frog, close-up view

Drab Tree Frog, concealed on a leaf

Red-eyed Stream Frog. Photo © Brian Kubicki

Duellmanohyla rufioculis

Total length: 1.2–1.6 inches (30–40 millimeters).

Range: Endemic to Costa Rica.

Elevational range: 1,450–5,000 feet.

RED-EYED STREAM FROG

This species might be confused with a juvenile Red-eyed Tree Frog (*Agalychnis callidryas*), but it is distinguished from that frog by the orientation of its pupil. Its pupil is horizontally elliptical, as opposed to the vertically elliptical pupil of *A. callidryas*.

The most common back color of this species is brown, with numerous small green spots, but on some specimens the back is a nearly uniform green. The belly is white. The white belly is one of the most obvious characteristics that separates it from the other two *Duellmanohyla* species in Costa Rica (*D. uranochroa* and *D. lythrodes*).

This species occurs along the middle elevations (premontane) of both the Caribbean and southern Pacific slopes of Costa Rica, where it is endemic. The Red-eyed Stream Frog is found near small headwater streams or seepages in humid forests. Males call from vegetation, branches, and rocks near or overhanging small streams and seeps during the evening. It is not known where this species deposits its eggs, but most likely they are attached to rocks and other submerged surfaces. The tadpoles are brown and have suckerlike mouthparts that allow them to adhere to rocks and other submerged surfaces in streams. Tadpoles can often be found in the same small pools where adults are seen.

One of the few places where it is possible to see this small uncommon species is the Costa Rican Amphibian Research Center near Guayacán.

LANCASTER'S SEEPAGE FROG

This small attractive frog varies in the color of its back. It may be a mixture of gray and brown, but in the Caribbean foothills near Siquirres it usually has more green on the back. The belly is white. The dorsal surface of the thighs has yellow spots and bands.

This uncommon frog is found mainly along middle elevation Caribbean slopes of the Talamanca Mountains, but some populations are known on the Caribbean slopes of the Turrialba Volcano. Males call from vegetation, rocks, roots, and branches overhanging small streams and seepages during the evening in humid forests. The eggs are deposited as a small patch of surface film attached to rocks and the edges of small pools in seepages and on side channels of streams. The tadpoles are tan with darker brown blotches, and they are often seen on the bottom of small pools. This beautiful and uncommon small frog can be seen at the Costa Rican Amphibian Research Center.

Isthmohyla lancasteri (formerly Hyla lancasteri)

Total length: 1.0–1.6 inches (30–40 millimeters).

Range: Regional endemic, known to inhabit a narrow altitudinal band along the Caribbean slopes of Talamanca (southeastern Costa Rica into northwestern Panama).

Elevational range: 1,320–4,000 feet.

Lancaster's Seepage Frog. Photo © Brian Kubicki

Green Stream Frog. Photo © Brian Kubicki

GREEN STREAM FROG

The first Green Stream Frog found in Costa Rica in the 1980s was originally considered a new species of glass frog, but later it was discovered to be *Hyloscirtus palmeri*. A small to medium-sized frog, its internal organs are visible through the translucent skin on the belly. There is a yellowish-white fringe along the outer edge of the lower arm and the lower leg and foot and a small yellowish-white calcar (spur) on the heel. Numerous small white spots are often present on the back. The iris ranges from silvery white to gray. The iris color depends on the activity level. Active or calling individuals normally have a darker iris.

Hyloscirtus palmeri (formerly Hyla palmeri)

Total length: 1.5–2.0 inches (35–50 millimeters).

Range: Costa Rica to Ecuador.

Elevational range: 1,650–3,000 feet.

The color of this frog can vary. The back can be olive-green to bright lime-green, depending on the activity level and temperature. While calling, the males often become olive. Blood vessels are easily seen through the translucent skin, and their abundance likely allows for increased oxygen exchange through the skin, given the aquatic calling habitat of the males. Males normally call from under the rocks in a stream, where highly oxygenated water is abundant. The call consists of a loud "chirp" or "chirp-chirp." It is not known where these frogs deposit their eggs, but most likely they are attached to submerged rocks or gravel in flowing streams.

This rare species is known to currently exist in Costa Rica only in the foothills near Siquirres, especially in the headwaters region of the Siquirres River. It is commonly heard calling along the Siquirres River at the Costa Rican Amphibian Research Center.

VAILLANT'S FROG

Lithobates vaillanti (formerly Rana vaillanti, formerly Rana palmipes)

Total length: 2.4–4.9 inches (67–125 millimeters).

Range: Veracruz, Mexico, to northern Colombia.

Elevational range: Sea level to 6,300 feet.

Costa Rica has five members of the leopard frog family, including Vaillant's Frog. The former specific epithet, *palmipes*, suggests one of its most distinctive features, the full webbing between the toes. Vaillant's Frog can be spotted during the day or night. This large frog resembles the Green Frog of North America. It lives in quiet forest backwaters and can generally be found at the water's edge, sitting quietly and waiting for prey to come close. It eats just about anything that is smaller than itself: small fishes, birds, rodents, arthropods, and even other frogs. When disturbed, it leaps into the pond and lies quietly under the water, often under a piece of aquatic vegetation. This is a common prey species for raccoons, coatis, herons, and egrets.

Vaillant's Frog is found throughout lowland and premontane moist and wet forests of the Caribbean slope and along streams and wetlands in Guanacaste. A good place to see this species is at the Costa Rican Amphibian Research Center.

Vaillant's Frog. Photo © Brian Kubicki

Ctenosaur and tourist Craig Henderson

REPTILES

Costa Rica's fauna includes at least 228 reptiles: lizards, geckos, skinks, snakes, marine and freshwater turtles, and crocodilians. Reptiles demonstrate impressive adaptations to the diverse tropical habitats of Costa Rica. They may be encountered from high mountains to coastal rainforests and in the dry forests of Guanacaste. Some are readily seen in the daytime, but others are best seen on guided night hikes. Many lizards depend on sunshine to warm their bodies, so iguanas, ctenosaurs, and *Norops* and *Ameiva* lizards are readily observed during the day. Along lowland streams, freshwater turtles and *Basiliscus* lizards can be seen basking along the shore.

Caimans can be observed during daytime boat trips along the canals of Tortuguero NP, along the Río Frío, and at La Laguna del Lagarto. Crocodiles are best seen from the Río Tárcoles bridge by the entrance to Carara NP, in Tortuguero NP, at La Ensenada Lodge, and at Estero Madrigal on Hacienda Solimar. These reptilians are prehistoric reminders of how long these creatures have inhabited the earth and of how important these tropical habitats are in perpetuating their existence.

An excursion to see nesting marine turtles on a peaceful moonlit beach is an unforgettable lifetime experience. Nocturnal guided tours to observe nesting Leatherback and Green Turtles are offered at Las Baulas NP (January to February) and Tortuguero NP (June to November), respectively.

Many people breathe a sigh of relief when they learn that snakes are seldom seen, even by enthusiastic naturalists who are intentionally looking for them. Most are nocturnal and live underground or in the treetops, where they are inaccessible. Although a few species like the bushmaster generate fear and misunderstanding, the danger posed by venomous reptiles is far less than that posed by aggressive drivers on Costa Rican roads. Snakes are seldom encountered and are harmless if left alone. Several very venomous snakes have been included among the species accounts because of the keen interest that people have in these snakes, but they are extremely rare and not likely to be encountered on a Costa Rican trip.

In all, forty-five reptiles have been selected for coverage here, including lizards, geckos, skinks, snakes, marine and freshwater turtles, and two crocodilians.

Ctenosaur, sunning on a stump

DOUBLE-CRESTED BASILISK (JESUS CHRIST LIZARD)

Double-crested Basilisk adult

Basiliscus plumifrons
Costa Rican names: *Chisbala*; *chirbala*.
9/23 trips; 18 sightings.
Total length: Up to 36.2 inches.
Range: Honduras to Panama.
Elevational range: Sea level to 2,325 feet.

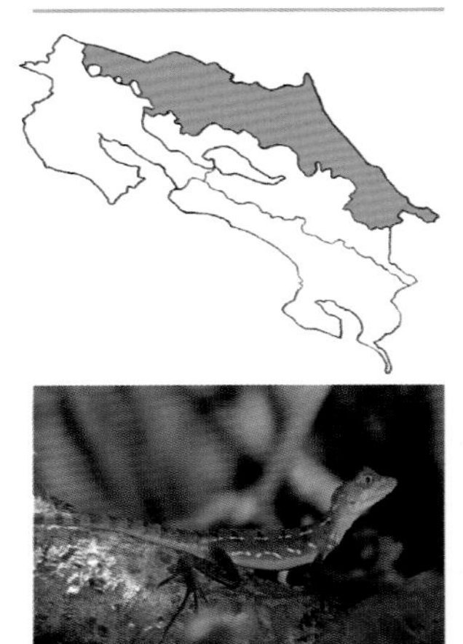

Double-crested Basilisk young

The memorable Jesus Christ Lizard is renowned for its ability to run across the water upright on its hind legs. A young basilisk can run more than fifty feet across a stream or pond if it drops from its perch on a branch overhanging the water.

There are three *Basiliscus* species, including this bright green, double-crested lizard in the Caribbean lowlands and southwestern Costa Rica. The prominent crest of *Basiliscus plumifrons* gives it the look of a miniature dinosaur. These crests aid in dissipating body heat in hot weather.

Basiliscus lizards lay five to eight clutches of eggs during the ten-month breeding season from April through January. Each clutch contains six to eighteen eggs that hatch after sixty to ninety days. Females become sexually mature at eighteen months of age. These lizards may live three to six years and continue to grow throughout their lives. These omnivorous lizards are active during the day and eat insects, newly hatched iguanas and ctenosaurs, small lizards, snakes, small birds, small mammals, fish, flowers, and fruits. Predators include snakes, hawks, and opossums, but the ability to run across water makes basilisks difficult to catch.

Look carefully for these well-camouflaged lizards resting on small branches overhanging streams and ponds. One way to discover them is to watch for their long, slender tails hanging from these branches, usually no more than one or two feet above the water. Look for them on ponds and canals in Tortuguero, La Selva, Santa Rosa, and Palo Verde NPs in Guanacaste; on streams of the Río Frío and Caño Negro NWR; and on pond edges in Corcovado NP on the Osa Peninsula. In good habitat they can be very common, at densities of 80–160 per acre.

Young Double-crested Basilisk, resting over water, where an Orange-eared Slider basks on a Black River Turtle

Double-crested Basilisk, resting along a stream

COMMON BASILISK (JESUS CHRIST LIZARD)

Basiliscus basiliscus

Costa Rican names: *Chisbala*; *chirbala*.

9/23 trips; 17 sightings.

Total length: Up to 35.4 inches.

Range: Tamaulipas, Mexico, to northern Colombia.

Elevational range: Sea level to 3,600 feet.

The Common Basilisk is found in the Pacific lowlands from Guanacaste south through the Osa Peninsula. It is brown with a prominent crest similar to the crest of the Double-crested Basilisk. The ranges of the two species overlap in coastal areas of southwestern Costa Rica, but the bright green Double-crested is readily distinguished from the brown species.

The Common Basilisk is found along wetland, coastal, and riverbank habitats in lowland and middle elevations of the Guanacaste dry forest, as well as in moist and wet forests of southwestern Costa Rica.

These large diurnal lizards are conspicuous as they rest on vegetation overhanging the water at wetland edges. Younger specimens of this lizard are famous for their ability to drop from a perch onto the water and run for more than fifty feet to escape from a potential predator. Adults are able to run only a few feet over the water before they sink and then swim to shore. The feet of this lizard are large, with long toes that are flattened at the edges to provide extra surface area. That allows them to run on the water without breaking the surface tension.

An omnivorous species, the Common Basilisk eats fruits and flowers, small invertebrates, fish, lizards (young iguanas and ctenosaurs), snakes, and small birds. Juveniles eat insects and small fishes. Predators include hawks, caracaras, opossums, and snakes.

The mating season occurs over ten months, from March through December. During that period, females lay up to eight clutches of eggs, and each clutch contains two to eighteen eggs. Offspring reach sexual maturity at twenty months of age, and individuals may live four to six years.

This lizard may be observed at La Pacífica, Villa Lapas, Carara NP, Rancho Casa Grande, Esquinas Rainforest Lodge, Hacienda Barú BR, Sirena Biological Station, and Corcovado NP.

Common Basilisk adult

Common Basilisk young

BROWN BASILISK (STRIPED BASILISK)

Basiliscus vittatus

Costa Rican names: *Chisbala*; *chirbala*.

5/23 trips; 8 sightings.

Total length: Up to 23.2 inches.

Range: Tamaulipas, Mexico, to Venezuela, Colombia, and Ecuador.

Elevational range: Sea level to about 360 feet.

The Brown Basilisk is a diurnal lizard that lives in the Caribbean lowlands. It is brown and has narrow white horizontal stripes through the face. It may be seen in habitats similar to those of the Double-crested Basilisk, but it is also found in upland sites away from water in the vicinity of brush piles and ground cover. Like other *Basiliscus* lizards, it can run across the water. Preferred habitats include lowland and middle elevations of the Caribbean slope.

The breeding season extends from March to October. Females may lay four or five clutches of two to eighteen eggs. Incubation requires fifty to seventy days. Young eat insects and arachnids. Adults are omnivorous and eat fruits, seeds, grasses, invertebrates and small vertebrates. Predators mainly include birds of prey. Adults live for two to three years.

This lizard may be found in the vicinity of Caribbean beaches, Tortuguero NP, and inland to La Selva Biological Field Station and the Sueño Azul resort.

Brown Basilisk, close-up view showing head detail

Brown Basilisk adult

Brown Basilisk young, running on water, the behavior that has earned basilisks the name "Jesus Christ Lizard"

CTENOSAUR (SPINE-TAILED LIZARD; BLACK IGUANA)

This ctenosaur is the most abundant and conspicuous large lizard on the Pacific slope of Costa Rica. On sunny days it spends most of its time basking on fence posts, rock piles, roofs, and tree limbs. The body is tan and dark brown, and there may be a deep reddish tinge on the back during the mating season. Juveniles are bright green.

This lizard could be mistaken for the Green Iguana, but that species is more greenish, has a longer tail, and has a large circular scale on the jaw area behind the eye. Iguanas are usually found by water, but ctenosaurs occur throughout the landscape on ranches, in backyards, in pastures, and in gardens. The tail of the ctenosaur is ringed by rows of large, pointed scales, which are responsible for the name "Spine-tailed Lizard." Sometimes ctenosaurs are hunted as small game, and they, as well as iguanas, are known as the "chicken of the tree," in reference to the similarity of their meat to chicken.

Although most abundant in the dry forests of Guanacaste, this lizard is found along the entire Pacific coast to the moist and wet forests of the Osa Peninsula and adjacent Panama. Its diet consists of leaves, flowers, field crops, fruits, small rodents, lizards, eggs of lizards and birds, frogs, insects, and other invertebrates. The diet changes from largely animal food for young ctenosaurs to largely plant food for adults. Ctenosaurs are eaten by raccoons, coatis, raptors, and boa constrictors. As it does on most other lizards, the tail will grow back if it is broken off.

The breeding season begins in December, when males may be seen defending their territories with head-bobbing displays. Females lay a single clutch of about forty eggs in underground burrows. The young hatch from April through July. Females reach sexual maturity at the age of two. This adaptable lizard may be seen high in the treetops; on low shrubs, posts, and branches; or foraging on the ground. It is common at ranches, resorts, and beachfront hotels throughout Guanacaste. Ctenosaurs can be seen along trails at Manuel Antonio NP; at La Pacífica, La Ensenada Lodge, and Hacienda Solimar; and at Palo Verde NP.

Ctenosaur adult, resting on a tree trunk in Guanacaste.

Ctenosaura similis

Costa Rican names: *Garrobo; iguana negra.*

18/23 trips; 52 sightings.

Total length: Up to 40 inches (101.6 centimeters).

Weight: 1 pound 5 ounces–2 pounds 5 ounces (596–1,050 grams).

Range: Southern Mexico to Panama.

Elevational range: Sea level to 1,000 feet.

Ctenosaur adult at Manuel Antonio National Park

A rare amelanistic ctenosaur that lacks normal skin pigmentation

Green Iguana adult male, showing large scale on its cheek area

GREEN IGUANA

The dinosaur character of the Green Iguana adds a touch of primitive wildness to tropical forests as it peers from treetop foliage. It is a symbol of rainforests well known locally as the "chicken of the tree" because of its use for meat. An adult iguana can be green to gray, with rust-colored highlights. Large males are frequently seen perched high in a tree, exposed in the sunlight, bobbing their heads in territorial displays. A large dewlap under the chin flaps conspicuously during the head bobs. This dewlap is absent on ctenosaurs. There is also a large, circular scale on the jaw area that is absent on ctenosaurs. The iguana tail is longer and more slender than the ctenosaur's, with alternating bands of grayish green and dark gray to black.

The iguana usually occupies sites overlooking rivers, canals, and wetlands. An adult male maintains a territory of about 0.2 acre, and females and young maintain territories of 0.5–0.6 acre. The iguana is a vegetarian throughout life, eating flowers, newly sprouted leaves, and fruits. When the mating season begins in December, male iguanas turn

Iguana iguana

Costa Rican name: *Iguana.*

21/23 trips; 74 sightings.

Total length: Up to 82.7 inches (210 centimeters).

Weight: Up to 15 pounds (6.8 kilograms).

Range: Northern Mexico to Paraguay and southern Brazil.

Elevational range: Sea level to 1,000 feet.

Green Iguana adult male, showing green color

bright reddish orange, and head-bobbing displays become conspicuous. A male may mate with one to four females in his territory. After mating, each female lays twenty-four to seventy-two eggs in an underground burrow. The young hatch after ten to fourteen weeks, in late April through May. At two to three years of age, iguanas reach sexual maturity. They may live up to fifteen years.

Young Green Iguana

Research and educational efforts have been under way to promote saving tropical rainforests by raising Green Iguanas for meat. This is a land-use alternative for local farmers who would otherwise cut the rainforest to raise cattle. One acre of rainforest can produce 300 pounds of iguana meat per year, but if cleared, it will produce only 33 pounds of beef per year.

Look for iguanas along the canals of Tortuguero NP, from the Stone Bridge over the Río Sarapiquí at La Selva Biological Field Station, from the bridge over the river in the town of Muelle, along the Río Tempisque and its tributaries in Guanacaste (including Palo Verde, Santa Rosa, and Guanacaste NPs), and in rainforest habitats of the southern Pacific lowlands, including Corcovado NP and Tiskita Jungle Lodge. The best place to see big iguanas up close is at the picnic area near the headquarters of the OTS field station at Palo Verde NP.

Green Iguana adult male in breeding condition

The dewlap displays of anoles, like this *Norops cupreus*, are delightful to watch.

The family Polychrotidae was designated in 1987 when more than 650 species of these small slender lizards were separated from the family Iguanidae. A subfamily within the Polychrotidae includes at least 189 species in the genus *Norops*. These lizards were formerly included in the genus *Anolis*. The lizards of that genus are still collectively referred to as anoles.

There are twenty-one species of *Norops* lizards in Costa Rica. (The giant anoles of the genus *Dactyloa* are related forms in the same family.) Most have brownish markings, but some females have prominent white stripes or diamond-shaped markings on the back. Their most conspicuous feature is a colorful dewlap on the throat that males extend as a territorial display against other males and to attract females. The dewlap color varies among species. Females may or may not have a dewlap, but on those that do, it is smaller than on the male and usually a different color. These diurnal and arboreal lizards are conspicuous on low shrubs, tree trunks, fence posts, and wood piles, as males extend and retract their dewlaps and bob their heads up and down.

Some anoles are able to regulate their body temperature and therefore can inhabit shady understory environments. Others need to bask and display in the sun to increase and maintain their body temperature.

Some *Norops* species breed year-round, producing about one egg per week during the rainy season and one egg every two to three weeks during the dry season. Young hatch after fifty days and become sexually mature

Anoles, found throughout much of Costa Rica, add an elfin presence to the vegetation.

in three to four months. If these small lizards are attacked by a predator, the tail will break off and twitch, distracting the predator and allowing the lizard to escape. It is common to see a distinct difference in color where a new tail has grown back on these lizards. The predators of anoles are numerous: praying mantises, katydids, trogons, motmots, pygmy-owls, screech-owls, Swallow-tailed Kites, orb-weaver spiders, vine snakes, and other lizards, including larger anoles. *Norops* lizards eat butterfly and moth caterpillars, crickets, katydids, grasshoppers, cockroaches, and other anoles. The Stream Anole even eats small fish.

These small lizards can be encountered throughout the country in many of Costa Rica's lowland and middle-elevation forests, plantations, ranches, and farms. The few species that are found at higher elevations in lower montane forests include *Norops vociferans*, *N. woodi*, *N. pachypus*, *N. tropidolepis*, and *N. altae*.

When trying to identify anoles, it can be useful to take a photograph for later comparison with a field guide. Some are difficult to identify, even in hand, however; but they are fascinating to watch as they display to defend their territories and attract females. It may be best simply to enjoy them as a component of their tropical habitat and not worry too much about determining the species.

One anole, the Golfo Dulce Anole, is endemic to southwestern Costa Rica. Nine species of *Norops* are included in these accounts. The author wishes to extend his appreciation to noted herpetologist Dr. Gunther Koehler for assisting with the identification of these *Norops* photographs.

A *Norops* lizard, caught by a Costa Rican Pygmy-Owl

DRY FOREST ANOLE

Norops cupreus

Total length: Up to 6.7 inches (17 centimeters).

Range: Eastern Honduras to central and southwestern Costa Rica to Quepos.

Elevational range: Sea level to 4,300 feet.

Norops cupreus is one of the most commonly encountered anoles in the Guanacaste region and in plantations, backyards, and gardens of the Central Plateau. This is a small gray to brownish lizard with a distinctive dewlap. The forward portion is pink and the rear portion is bright orange. Within Guanacaste, this lizard is most common in gallery forests and in forested beachfront areas. The range extends south along the coast to the Hacienda Barú region.

Surveys have counted up to 770 Dry Forest Anoles per acre. These lizards maintain territories that they defend against others of the same species. Most foraging for ants, caterpillars, sowbugs, cockroaches, ticks, cicadas, and spiders occurs in leaf litter, among low shrubs, on fences, and in rock piles. This is one of the few anoles in Costa Rica that is adapted to the hot, dry environments of Guanacaste. In late April, females begin to lay an egg about every week, and they continue to lay through October. Young begin hatching in July, and they reach sexual maturity in one year.

This is an abundant ground-level anole, so it is one of the first anoles that a visitor to Guanacaste is likely to encounter, probably on a fence post.

Dry Forest Anole

Dry Forest Anole male, displaying

GROUND ANOLE

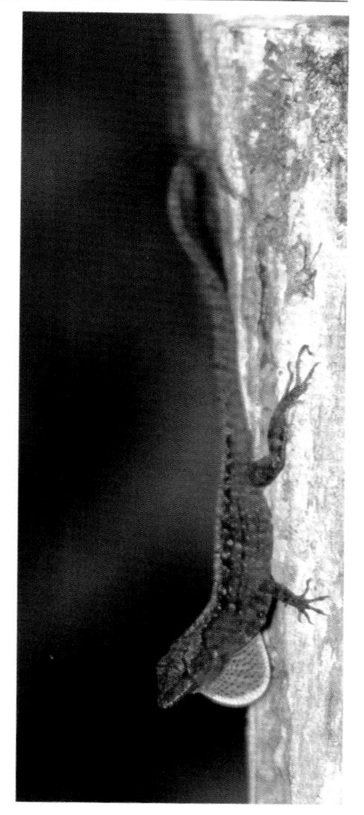
Ground Anole, displaying red and yellow dewlap

The Ground Anole is aptly named because it spends its life on or within a couple feet of the ground. The body is brownish with darker brown markings along the back. Many anoles spend a significant part of the day in the sun to warm their body, but this species spends its life in the shade and can regulate its body temperature without basking in the sun. The body of this lizard is less than two inches long. Identifying characteristics include a pocket in each armpit of the forelegs; no function is known for the pockets except that they are typically occupied by tiny mites or ticks. There are also eight to ten rows of scales down the middle of the back that are significantly larger than the scales on the remainder of the body. The dewlap of the male is orange or bright red with a conspicuous yellow edge.

The Ground Anole lives in leaf litter on the ground of lowland and premontane moist, wet, and rainforests. An adaptable species, it can inhabit both pristine forests and plantations.

A male will typically climb up a couple of feet on the buttresses of rainforest trees, head downward, either displaying or waiting for prey species to approach. Prey includes spiders, sowbugs, beetles, caterpillars, flies, termites, and centipedes.

Reproduction occurs primarily during the rainy season. Females lay about one egg per week in leaf litter. The eggs hatch between forty-six and fifty-seven days. Most anoles do not live more than a year.

This is a common lizard in the Caribbean lowlands. Densities can approach 200–450 per acre. It is encountered in places like La Selva Biological Field Station.

Norops humilis

Total length: Up to 4.5 inches (11.4 centimeters).

Range: Central Costa Rica to Panama.

Elevational range: Sea level to 2,900 feet.

Pug-nosed Anole, with impressive camouflage

PUG-NOSED ANOLE

This distinctive anole is readily identified by its short head and short, prominent snout. The legs are very long on this common lowland rainforest species. The markings and body colors are highly variable, as they are among many anoles. It may be plain or mottled with yellowish green, greenish, rusty brown, or olive-brownish, with several darker diagonal bands along the sides. The coloration provides excellent camouflage as the anole clings to tree trunks. The body, exclusive of the tail, may be up to three and a half inches long. The dewlap is greenish yellow.

This arboreal species may be found at heights above fifty feet in trees, but they are usually within six feet of the ground. Like the Ground Anole, this lizard lives in the shade rather than in the sun and is able to regulate its body temperature without the warming effects of the sun.

Prey species include spiders, ants, flies, katydids, caterpillars, and slugs. Being large for an anole, it is aggressive and will sometimes capture and eat smaller anoles.

Habitat includes relatively undisturbed moist and wet tropical lowlands and premontane forests on both the Caribbean slope and the southern Pacific slope. One location where it can be encountered is La Selva Biological Field Station.

Norops capito

Total length: Up to 10.5 inches (26.6 centimeters).

Range: Atlantic slope from Mexico to Panama; Pacific slope from southwestern Costa Rica to Panama.

Elevational range: Sea level to 3,600 feet.

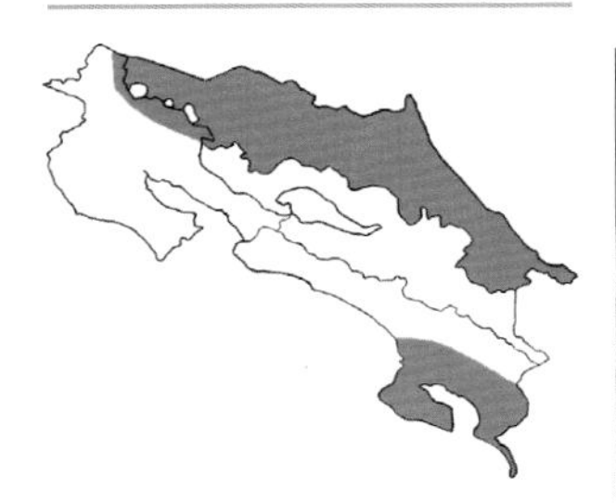

Pug-nosed Anole on a biologist's hand for size comparison

GOLFO DULCE ANOLE

Norops polylepis

Total length: Up to 7 inches (17.8 centimeters).

Range: Southern Pacific lowlands of Costa Rica and western Panama.

Elevational range: Sea level to 4,000 feet.

The Golfo Dulce Anole is a common *Norops* lizard of Costa Rica's southern Pacific lowlands. It is easily observed because it forages and conspicuously displays on the ground and on low shrubs. The coloration can vary from a uniform creamy tan or brown to medium brown, reddish brown, or yellowish brown. Males usually have more uniform coloration, and females can have diamond-shaped patterning along the back.

Males defend their territories from perches near the ground by bobbing their heads and extending their dewlaps to deter other males and to attract females. The dewlaps are bright orange. This anole prefers understory habitat in the shade rather than open areas in the sun.

This lizard lives a fast life. It becomes reproductively mature in only three to four months. Females will mate and lay about one egg per week. Each egg hatches after an incubation period of about fifty days. The lifespan is generally not more than a year.

Clinging to a tree trunk or branch with head downward, the Golfo Dulce Anole waits for the approach of prey such as katydids, caterpillars, and other small invertebrates. It serves as prey for snakes, mammals, raptors, trogons and motmots.

Look for the Golfo Dulce Anole in moist to wet lowlands and premontane moist to wet forests from Carara NP southward to the Panama border, including San Vito Biological Station and Manuel Antonio and Corcovado NPs.

Golfo Dulce Anole adult male, displaying

Golfo Dulce Anole, showing back markings

Golfo Dulce Anole, with tail growing back

SLENDER ANOLE

Norops limifrons

Total length: Up to 6.1 inches (15.6 centimeters).

Range: Eastern Honduras to central Panama.

Elevational range: Sea level to 4,000 feet.

As one would expect from the name, this anole has a very slender profile. It is grayish brown above and demonstrates considerable variation in markings. One of the most distinctive markings is the white upper lip. Some may have a cream-colored stripe down the center of the back with a slender black or dark line on each side of the stripe. Others have a marbled appearance, with a series of dark markings along the center of the back. The tail usually shows banding. The male's small dewlap is white with an orange to yellowish spot in the center.

This is a small, very common lizard in Costa Rica's tropical moist and wet lowland rainforests and in lower premontane forests. An adaptable species, it is found on both the Caribbean and southern Pacific slopes in undisturbed forests, second-growth forests, plantations, and even shrubby pastures.

The Slender Anole lives on the lower levels of tree trunks and shrubs, usually within six to seven feet of the ground, and it may also be found in ground cover. The species prefers shady areas and avoids displaying or hunting in direct sunlight. Prey includes spiders, mites, beetles, katydids, flies, and caterpillars. From a perch on a tree trunk, it may spot prey and chase it down.

One of the most remarkable behavioral aspects of this anole is its monogamous pair bonds. The male and female are usually encountered within several feet of each other, and they move through their territory together. The bonded pair will frequently do head bobs to each other. The female will lay an egg about every week during the rainy season and about once every three weeks during drier periods. After hatching, the young reach sexual maturity within fifty-eight to sixty days. Their life expectancy is less than a year. They are prey items for motmots, trogons, puffbirds, snakes, mammals, and even Golden Orb-Weaver spiders.

This *Norops* lizard can be encountered throughout the Caribbean lowlands, including La Selva Biological Field Station. It also occurs in the southern Pacific lowlands from Carara NP to the Panama border.

Slender Anole, showing a dorsal stripe

GREEN TREE ANOLE

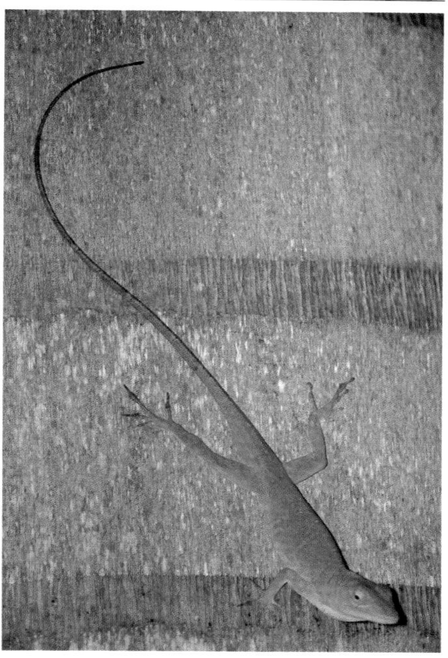

Green Tree Anole adult

This large, bright green lizard is typically seen clinging to a tree trunk. It has a long slender tail that may account for two-thirds of its total length. *Dactyloa frenata* is also a large green lizard in the family Polychrotidae, but it is rare, and the four diagonal bands between its hind and forelegs are absent on the Green Tree Anole. The dewlap on the male anoles is light bluish near the throat and reddish orange along the outer edge. This *Norops* species can undergo impressive color changes from green to splotchy brown.

A sun-loving lizard, it inhabits lowland and premontane levels of moist and wet rainforests throughout the Caribbean and southern Pacific slopes of Costa Rica. It is absent from dry forests and at high elevations. This diurnal species hunts for small invertebrates by clinging to tree trunks facing upward and watching for movements of beetles, spiders, large ants, and smaller *Norops* lizards.

Most sightings of the Green Tree Anole are on tree trunks within about ten feet of the ground, but they are adept climbers and have been encountered at heights up to 115 feet in the canopy, where they hunt for invertebrates or find locations for laying eggs. One egg is laid at a time, but reproductive details are poorly known. This feisty lizard can bite if captured, and it is one of the only *Norops* species that can squeak if picked up.

The *Norops biporcatus* specimen shown here was photographed at the Wilson Botanical Garden near San Vito.

Norops biporcatus

Total length: Up to 13 inches (33 centimeters).

Range: Mexico to Venezuela and Ecuador.

Elevational range: Sea level to 3,660 feet.

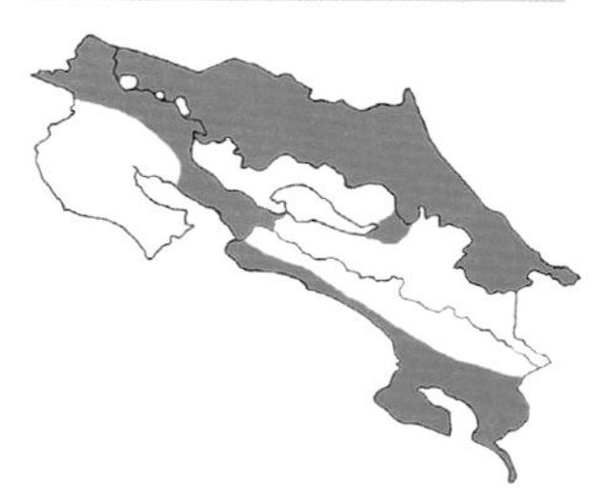

STREAM ANOLE

Norops oxylophus

Total length: Up to 9.7 inches (24.3 centimeters).

Range: From eastern Honduras to western Panama.

Elevational range: Sea level to 3,600 feet.

The Stream Anole is distinctively marked, with a whitish racing stripe along each side, and the snout has a prominent duck-billed profile. This lizard is brownish to olive over the back and cream-colored on the belly. The male's dewlap is yellowish orange. Although a few specimens have been encountered in the southern Pacific lowlands, it is more common in the Caribbean lowlands and at middle elevations of Costa Rica's mountains.

This shade-loving species is found in the forest understory along streams and rivers, where it may be observed on rocks and logs in water and on streamside vegetation. If approached, it will readily jump into the water to escape. It is an accomplished swimmer. In addition to eating a variety of invertebrates like spiders, flies, and beetles, it will also catch and eat small fish.

Females lay one egg at a time in mossy, moist leaf litter throughout the year. After hatching, the young become reproductively mature after only 101–118 days. Look for this anole along rivers in the vicinity of La Selva Biological Field Station.

Stream Anole, with conspicuous white line on the side, like a racing stripe

CANOPY ANOLE (GHOST ANOLE)

Norops lemurinus

Total length: Up to 9 inches (22.8 centimeters).

Range: Mexico to western Panama.

Elevational range: Sea level to 2,900 feet.

The Canopy Anole has another name, "Ghost Anole," because the female has skeletal-looking white markings along the back. Although this anole can be found on tree trunks near the ground, "Canopy" applies because it is extremely agile and often occurs in the rainforest canopy. The male is usually gray or brown on the back. One distinguishing mark is a W-shaped bony structure on the back of the head. The male has a small dark red dewlap. The female is one of the few anoles with a well-developed dewlap that is white. The belly and throat are white to cream-colored.

The distribution of the Canopy Anole includes lowlands and premontane forests of the Caribbean slope and, to a lesser extent, lowlands of the southern Pacific slope. Shaded locations are preferred within those forests.

Prey includes spiders, beetles, caterpillars, flies, and katydids. Breeding occurs primarily during the rainy season from May through December. As with other anoles, predators include motmots, trogons, puffbirds, mammals, and snakes. Since these lizards frequently inhabit foliage in the upper canopy of the rainforest, they are also taken by Swallow-tailed Kites that snatch them from the foliage while in flight.

Canopy Anole, showing skeleton markings on the back

LICHEN ANOLE

Norops pentaprion

Total length: Up to 7 inches (17.8 centimeters).

Range: Mexico to Colombia.

Elevational range: Sea level to 2,700 feet.

Lichen Anoles have a mottled gray and brown body that blends into the lichen on the trees they inhabit. The short, muscular tail is pale and ringed with darker brown bands. The tail likely is adapted to guide the body while gliding. The dewlap of the male is reddish purple; the female's dewlap is slightly smaller than the male's.

The Lichen Anole could be called the glider anole; it has a wide, flattened body and loose skin between the forelegs and body that make it possible for the anole to jump from a perch and glide to nearby trees or branches with its legs outspread, like a flying squirrel. Other anoles are extremely agile and may run or jump from one branch, but they do not have the ability to glide.

This anole is arboreal and diurnal. It spends most of its time in the top of the canopy, where it basks in the sun. The diet consists of flies, beetles, ants, and spiders.

This widely distributed lizard is found throughout dry, moist, and wet areas and rainforests in tropical lowlands on both the Caribbean and Pacific slopes of Costa Rica.

Lichen Anole, showing deep red dewlap with pale blue spots

Lichen Anole adult

GREEN SPINY LIZARD (MALACHITE LIZARD)

Sceloporus malachiticus
Costa Rican name: *Lagartija espinosa.*
16/23 trips; 28 sightings.
Total length: Up to 7 inches (17.8 centimeters).
Range: Veracruz, Mexico, to Panama.
Elevational range: 1,900–9,200 feet.

As this colorful spiny lizard basks in the midday sun, it becomes bright malachite green. The color darkens during cooler portions of the day, and at night it turns blackish. This adaptable high-elevation lizard inhabits rocky outcrops, exposed dirt banks along highways, garden walls, patios, and rural backyards. It usually clings to rocks, banks, or tree trunks in a vertical posture.

The diet includes insects, flowers, fruits, and plant sprouts. The spiny lizard has a unique adaptation whereby eggs develop and hatch inside the female. The soil is too cool at high elevations for the eggs to develop, but the heat needed for incubation is derived from the basking of the female. She gives birth to an average of six young each year, usually in January and February. Watch for this small but brightly colored lizard around garden and patio areas of the Central Plateau and highland farms, hotels, and nature resorts like those at Monteverde and San Gerardo de Dota.

Resplendent Quetzal eating a Green Spiny Lizard

Green Spiny Lizard, in typical basking posture

Green Spiny Lizard female, well camouflaged on a tree trunk

AMEIVA LIZARD (CENTRAL AMERICAN WHIPTAIL)

Ameiva festiva

Costa Rican name: *Chisbalas.*

10/23 trips; 11 sightings.

Total length: 8–9 inches (20.3–22.9 centimeters).

Range: Tabasco, Mexico, to Colombia.

Elevational range: Sea level to 4,900 feet.

This whiptail is a medium-sized speckled lizard of Costa Rica's Caribbean and southern Pacific lowlands. Except for large males, the species has a prominent light stripe down the center of the back and is conspicuous when sunning along rainforest trails. There are three other members of the *Ameiva* genus in the country: *A. quadrilineata* (Caribbean and southern Pacific lowlands), *A. undulata* (northwestern Costa Rica and the Central Plateau), and *A. leptophrys* (southwestern Costa Rica). *Ameiva festiva* is the most widely distributed. It feeds primarily on the ground at forest edges, where it digs and explores among leaves and twigs for small insects, insect eggs, and insect larvae.

A female produces three or four clutches of eggs each year, with an average of two or three eggs per nest. Predators include hawks, motmots, opossums, coatis, and other medium-sized carnivores. *Ameiva* lizards may be seen at La Selva Biological Field Station and Rancho Naturalista in the Caribbean lowlands and at Carara and Corcovado NPs in the southern Pacific lowlands.

Central American Whiptail, a common lizard of lowland rainforests

DEPPI'S WHIPTAIL

Cnemidophorus deppii
Costa Rican name: *Lagartija.*
1/23 trips; 1 sighting.
Total length: Up to 11.6 inches (29.5 centimeters).
Range: Mexico to northwestern Costa Rica.
Elevational range: Sea level to about 1,000 feet.

This colorful and sun-loving whip-tailed lizard is found in sandy areas with sparse vegetation in Guanacaste and south to the Puntarenas area. The body is highlighted by thin cream-colored longitudinal stripes, red sides, and bluish markings on the throat and belly. The tail is long and slender. Young have a very bright blue tail. It is frequently found near oceanfront beaches and sand dunes, where it searches for grasshoppers and digs for spiders and termites. It is a shy species that dashes for cover when approached.

Reproduction occurs primarily during the rainy season, from March through October. The female lays a clutch of about three eggs, which hatch after sixty days. Four to six clutches may be laid during a year.

Look for this beautiful lizard along the beaches of Guanacaste and on sandy areas of sparse cover on ranches, backyards, and farms. The example shown here was photographed in the courtyard of La Ensenada Lodge.

Deppi's Whiptail, in Guanacaste

YELLOW-HEADED GECKO

Gonatodes albogularis

Costa Rican name: *Lagartija.*

Total length: Up to 4.4 inches (11.3 centimeters).

Range: Mexico to South America.

Elevational range: Sea level to 500 feet.

Unlike most geckos, which are active at night, the Yellow-headed Gecko is a conspicuous and diurnal species. The male has a distinctive bright yellowish to orange head, and the tail (if not regrowing) is about the length of the body. It has slender toes and small pointed toenails instead of the large blunt toes characteristic of most nocturnal geckos. Another difference is the large eyes with round pupils, which are an adaptation for hunting in daylight. An inhabitant of shaded areas in forest understory, back-yards, gardens, outbuildings, and fencelines, this gecko hunts for small invertebrates like spiders.

The Yellow-headed Gecko breeds throughout the year, but there is a peak of reproduction during the rainy season. The female lays one egg, which hatches after four months. The young become sexually mature after six months. Males are very territorial against other males. If approached by another male, a gecko will carry out a dramatic exhibition of arching and twitching its tail over its back, raising up its body, and jerking its head. Colonies of these geckos may live in large strangler fig (*Ficus*) trees.

The skin of this gecko is fragile, and the tail is easily broken off if grabbed by a person or predator. To avoid injury to or loss of the tail, it is suggested that lizards not be caught for casual examination. Insect repellent on a person's hands can also cause injury to an amphibian or reptile.

This adaptable lizard is found in dry, moist, and wet lowlands along the Pacific and Caribbean slopes from Nicaragua to Panama, but it is most abundant in the Guanacaste dry forest region.

Yellow-headed Gecko female, with beginning of a new tail

HOUSE GECKO

Hemidactylus frenatus

Costa Rican name: *Lagartija.*

Total length: Up to 5.3 inches (13.5 centimeters).

Range: From the Old World tropics, introduced in the American tropics.

Elevational range: Sea level to 1,000 feet; mostly coastal regions on both Caribbean and Pacific slopes.

The common House Gecko is an adaptable lizard and worldwide traveler that has hitch-hiked its way around the world from its Asian origins to New Guinea, Hawaii, East Africa, Florida, Texas, Mexico, Madagascar, and Central America. This gecko was first recorded in Costa Rica in the early 1990s. A very noisy little gecko that often inhabits tourist cabins in coastal areas, it makes loud barking "chacks" through much of the night, much to the chagrin of sleepy tourists.

This small aggressive gecko stalks insects, such as moths that have been attracted to security lights at cabins. It can be active during the day but is mostly active at night. The body color is brownish to grayish during the day and becomes pale at night.

Reproduction occurs through much of the year. A female usually lays two eggs in a clutch, in nooks and crannies of houses and outbuildings. Eggs hatch in forty-five to ninety days. One interesting feature of this gecko is that the females can retain viable sperm for up to eight months and can continue to produce up to ten clutches of eggs from that stored sperm.

The gecko shown here was photographed at La Pacífica in Guanacaste. You don't need to look for this gecko; it will find you as you are settling into your cabin after a day afield.

House Gecko, stalking a moth

MOURNING GECKO

Lepidodactylus lugubris
1/23 trips; 1 sighting.
Total length: Up to 4.1 inches (10.5 centimeters).
Range: From Asia, introduced to many countries.
Elevational range: Sea level and adjacent coastal areas.

This gecko is another adaptable exotic species like *Hemidactylus frenatus*. It has been accidentally introduced from Asia to seaport communities in many other countries, including New Zealand, Florida, Central America, South America, and islands of the South Pacific.

Acrobatic and adaptable, this gecko frequents nooks and crannies of homes and cabins in southwestern Costa Rican coastal areas, from Quepos to the Osa Peninsula, and adjacent areas of the Golfo Dulce lowlands. It has huge pads on its toes, like the classic gecko feet, sharp little claws on all but the first toe, and webs between the toes. It is a nocturnal species that frequents areas around security lights, where it pursues insects attracted to the lights. The gecko shown here was photographed at La Cusinga Lodge near Playa Dominical.

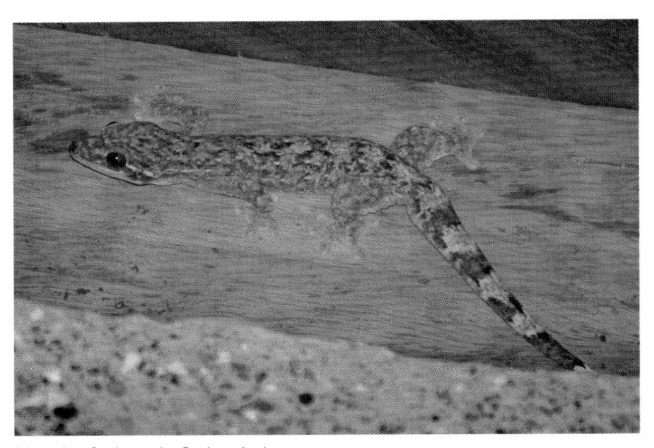

Mourning Gecko, at La Cusinga Lodge

TURNIP-TAILED GECKO

Thecadactylus rapicaudus
1/23 trips; 1 sighting.
Total length: Up to 8.9 inches (22.5 centimeters).
Range: Central America, northern South America, and Caribbean islands.
Elevational range: Sea level to 3,450 feet.

This is the largest gecko in Costa Rica, approaching nine inches in length. The broad tail looks swollen, in the shape of a turnip, because it is used for fat storage. An arboreal and nocturnal species, it inhabits lowland and premontane moist and wet forests on both the Caribbean and southwestern Pacific slopes. Sometimes it can be seen during the day on or near the ground on trunks of large trees or tree buttresses, but it normally spends daytime hours under the loose bark of trees or in tree crevices. It may also inhabit the thatching of rainforest houses and outbuildings, where it comes out at night on the ceilings and walls to catch insects. Primarily nocturnal, it can sometimes be discovered clinging to the trunks of trees, where it hunts for katydids and cockroaches.

Like other geckos, it has well-developed pads, or clinging lamellae, on the toes. This helps the gecko cling to tree bark as it pursues small insects and other invertebrates at night. The yellowish-brown to dark brown markings help camouflage geckos on a background of tree bark. If approached by a predator, this gecko arches its back like an upset cat, opens its mouth, and displays a conspicuous blue tongue. It can inflict a painful bite. If attacked, the tail readily breaks off to distract the predator, and a new one grows back.

Look for the Turnip-tailed Gecko on rainforest tree trunks while on guided night walks. (Do not go out wandering in the forest at night by yourself.) The gecko shown here was encountered along the loop nature trail at Tortuga Lodge.

Turnip-tailed Gecko, at Tortuga Lodge

BRONZE-BACKED CLIMBING SKINK

Mabuya unimarginata

Costa Rican name: *Lagartija.*

1/23 trips; 1 sighting.

Total length: Up to 9.7 inches (24.6 centimeters).

Range: Central Mexico to Panama.

Elevational range: Sea level to 4,950 feet.

As the name suggests, this skink has a bronze-colored back; there is also a dark band along the side with a narrow white line above and below. This skink has a transparent lower eyelid that allows it to see when its eyes are closed.

An adaptable and diurnal species, this skink is found in habitats varying from the dry forests of Guanacaste to moist and wet lowland forests on both the Caribbean and Pacific slopes and up to lower montane forests. It can be found in natural forests as well as in banana plantations, and along fencelines and roadsides. A shy species, it is very hard to approach. The diet consists of small arthropods such as katydids, spiders, and grasshoppers. One of the most notable features of this skink is that it is viviparous. It gives birth to four to seven live young instead of laying eggs. The skink shown here was photographed on a fenceline along a road in Guanacaste.

Bronze-backed Climbing Skink, in Guanacaste

NEOTROPICAL SUNBEAM SNAKE

This interesting snake is closely related to the boa constrictor. It is black, bluish, gray, or dark brown, with a glossy iridescent appearance, and the belly and chin are cream to yellowish in color. The name is apparently derived from the iridescent qualities of the body. A nocturnal egg-eating specialist of the Guanacaste region, it is found in very dry, sandy habitats in dry forests, on ranchlands, and along beaches, where it searches for nests and eggs of ctenosaurs, iguanas, and Ridley sea turtles. The upturned nose is used to dig up and expose eggs in nests. Other incidental prey includes small ctenosaurs and small mammals.

When eating sea turtle eggs, the sunbeam snake will grip the egg with its body, constrictor fashion, and then swallow the egg. The snake shown here was photographed at night on Playa Grande near Tamarindo in 1989 in an area being used by Leatherback Turtles for nesting. The snake was in the vicinity of a nest of emerging Leatherbacks, but no predation was observed.

Loxocemus bicolor

1/23 trips; 1 sighting.

Total length: Up to 60.2 inches (153 centimeters).

Range: Mexico to northwestern Costa Rica.

Elevational range: Sea level to 300 feet.

Neotropical Sunbeam Snake, at a Leatherback Turtle nest

Boa constrictor

BOA CONSTRICTOR

The boa constrictor is the largest snake in Costa Rica. Although the prospect of seeing a snake that can exceed ten feet in length might create anxiety, this reptile is not commonly seen. It feeds primarily on small mammals and birds and is not a threat to people. An adaptable snake, the boa inhabits dry and riparian forests in Guanacaste and lowland and middle-elevation forests of the Caribbean and southern Pacific regions. It occurs in pristine forests as well as in cleared areas and on farms and ranches. Sometimes one will take up residence in the rafters of a building with a thatched roof, where it helps control rodents at night, rather like a six-foot-long cat with scales.

As the name "constrictor" implies, this reptile kills its prey by coiling around it, suffocating it, and swallowing it headfirst. Prey includes small deer, coatis, raccoons, tamanduas, agoutis, tepescuintles, bats, ocelots, lizards, and birds. Dogs and poultry may also be taken. The boa hunts primarily at night and at dawn and dusk. It will remain motionless on a tree branch or in an animal's burrow and await the passage of a bird or mammal.

This snake gives birth to twenty to sixty-four live young from March through August. The young are about eighteen inches long at birth and disperse with no subsequent parental care. A boa becomes sexually mature at a length of five to six feet. This snake may live more than thirty years, but in the past it has been killed indiscriminately by people for its hide and by people who fear snakes.

Boa constrictor

Costa Rican names: *Boa; béquer.*

7/23 trips; 8 sightings.

Total length: Up to 15 feet (457.2 centimeters); record is 19.7 feet (600.4 centimeters).

Weight: Up to 66 pounds 2 ounces–88 pounds 3 ounces (30–40 kilograms).

Range: Sonora and Tamaulipas, Mexico, to Argentina and Paraguay.

Elevational range: Sea level to 3,800 feet.

Boa constrictor, in Rincón de la Vieja National Park

Boa constrictor, with captured ctenosaur, beginning to swallow it headfirst

CHUNK-HEADED SNAKE

Imantodes cenchoa

Costa Rican name: *Bejuquilla.*

1/23 trips; 1 sighting.

Total length: Up to 49.2 inches (125 centimeters).

Range: Mexico to Argentina.

Elevational range: Sea level to 4,900 feet.

This arboreal snake is named for its large head, which appears to be out of proportion to its pencil-thin body. The body is brown with dark saddlelike spots along the back. The eyes have distinctive vertical pupils, like those of a cat. A nocturnal species, it is adapted to climbing out on the tips of tree branches and slender twigs in search of anoles (*Dactyloa* and *Norops*) and small *Eleutherodactylus* frogs that sleep on foliage. During the day, this snake typically rests in bromeliads or under loose tree bark.

The cross-section of the body of this snake is shaped like an I-beam, providing great structural strength that allows it to extend its body from one branch to another out to half of its total body length. The snake can approach its prey with great stealth because of this. Once a victim like a *Norops* lizard has been caught, the fangs in the back of the snake's mouth inject a mild venom that incapacitates the lizard.

The Chunk-headed Snake reaches sexual maturity at two years of age. Females typically lay two or three eggs, and the breeding season extends through the year.

This is a common, gentle snake that presents no threat to humans. It does not attempt to bite. It is distributed throughout most of Costa Rica, except for high-elevation montane forests and the dry forests of Guanacaste. Preferred habitats include primary and secondary moist and wet forests as well as coffee and banana plantations. The only way to encounter this snake is to look for it on trailside vegetation on night walks that are led by professional naturalist guides. The snake shown here was found on the loop trail behind Tortuga Lodge near Tortuguero NP.

Chunk-headed Snake, a predator of tree frogs

Vine snake *Oxybelis brevirostris*

Oxybelis species
Costa Rican name: *Bejuquillo.*
2/23 trips; 6 sightings.
Total length: Up to 72 inches (182.9 centimeters).
Range: Mexico to Brazil.
Elevational range: Sea level to 2,600 feet.

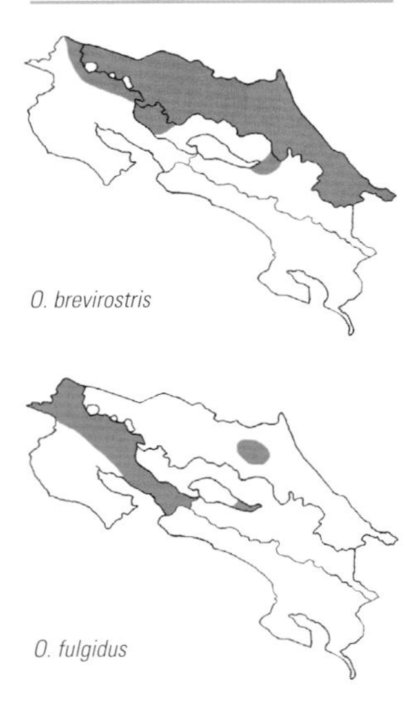
O. brevirostris

O. fulgidus

VINE SNAKES

The pencil-thin profile of these snakes serves as excellent camouflage; as they drape themselves over thin branches, they mimic a slender vine and wait for small lizards. They are difficult to discern in vegetation, so sightings are uncommon. The term "vine snake" is loosely applied to several long, slender arboreal snakes, including three species in the genus *Oxybelis: O. aeneus* (a gray snake found on the Pacific slope in dry forests of the north and humid forests of the south), *O. brevirostris* (shown here, a brown-green and white snake found in Caribbean lowlands, including Tortuguero NP and La Selva), and *O. fulgidus* (shown below, a green snake of the Guanacaste dry forest and northeastern Caribbean lowlands).

These snakes have an extremely slender body. The color is green, brownish, or grayish above and creamy white or greenish below. The head is long, slender, and pointed. The round pupil is black. If approached, *O. aeneus* opens its mouth as a bluff and exposes a black mouth lining. A rear-fanged snake, it can bite, chew, and inject a mild venom that may cause local blisters and swelling.

Vine snake *Oxybelis fulgidus*

GREEN-HEADED TREE SNAKE (PARROT SNAKE)

Leptophis ahaetulla
8/23 trips; 6 sightings.
Total length: Up to 88.6 inches (225 centimeters).
Range: Mexico to northern Argentina.
Elevational range: Sea level to 4,200 feet.

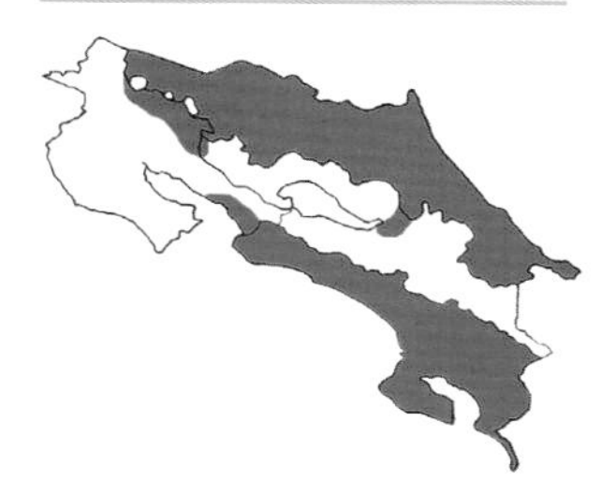

Snakes in the genus *Leptophis* are considered Green-headed Tree Snakes, including five species in Costa Rica: *L. ahaetulla* (described here) of Caribbean and southern Pacific humid lowland forests, *L. depressirostris* and *L. nebulosus* of Caribbean and southern Pacific lowlands, *L. mexicanus* of the Guanacaste dry forest and Caribbean lowlands, and *L. riveti* of southern Pacific lowlands and foothills. Rear-fanged, with a mild venom, these snakes feed primarily on tree frogs.

Leptophis ahaetulla is greenish above and pale yellow below. Short, black lateral stripes pass from the nose through the eyes to the back of the head, and the eyes have round, black pupils. They usually do not exceed 36 inches in length and do not pose a threat to people, but large specimens have been recorded up to 88 inches.

This is a diurnal and arboreal species that frequents areas of second-growth forest. Its primary prey consists of frogs and young birds. It breeds mainly in the rainy season and lays clutches of one to five eggs in the cover of bromeliads. During its interesting threat display, this snake will hiss while opening its mouth to expose a blue tongue sheath.

The best way to look for this snake is by carefully examining courtyard vegetation at tourism lodges of the Caribbean and southern Pacific lowlands, such as Tortuga Lodge, Esquinas Rainforest Lodge, and Drake Bay Wilderness Resort.

Green-headed Tree Snake

MEXICAN GREEN-HEADED TREE SNAKE

Leptophis mexicanus
8/23 trips; 6 sightings.
Total length: Up to 50 inches (126.9 centimeters).
Range: Mexico to Costa Rica.
Elevational range: Sea level to 3,480 feet.

This species of Green-headed Tree Snake (or Parrot Snake) has a green head, yellowish white sides, and black lateral stripes that extend the length of the body. The eyes have round, black pupils. Rear-fanged, with a mild venom, these snakes feed primarily on tree frogs. They usually do not exceed 36 inches in length (but may reach up to 50 inches) and do not pose a threat to people.

This diurnal tree snake is found in the dry forest region of Guanacaste, where it occurs mainly in gallery forests along streams and rivers. Besides tree frogs, it feeds on *Norops* lizards, small snakes, and bird eggs. Reproduction occurs mainly during the rainy season, from March through November. Clutches include two to six eggs.

Look for this snake in Santa Rosa and Palo Verde NPs, at Lomas Barbudal BR, and at streamside locations on ranches and lodging sites like La Pacífica, La Ensenada Lodge, and Hacienda Solimar.

Mexican Green-headed Tree Snake

RED COFFEE SNAKE

Ninia sebae

Costa Rican name: *Culebrilla de café.*

0/23 trips; 0 sightings.

Total length: Up to 15.2 inches (38.6 centimeters).

Range: Mexico to Costa Rica.

Elevational range: Sea level to about 3,300 feet.

The Red Coffee Snake is bright red, with a black head and two yellow bands behind the head. The only other snake that is mostly red is the juvenile *Clelia clelia,* which has only one light creamy yellow band behind the head. This harmless snake burrows through leaf litter in search of snails, leeches, worms, slugs, and small frogs. It occurs frequently on coffee plantations, thus the name.

A resident of primary forests as well as secondary forests, coffee plantations, farmlands, and pastures, this species is found mainly in northern regions of Costa Rica, in lowland and premontane moist and wet forests. Since it is usually concealed in leaf litter, it is seldom seen. The specimen shown here was photographed at Chan Chich Lodge in Belize.

Red Coffee Snake

FALSE FER-DE-LANCE

Xenodon rabdocephalus
1/23 trips; 1 sighting.
Total length: 26.8–35.4 inches (68–90 centimeters).
Range: Veracruz, Mexico, to Brazil.
Elevational range: Sea level to 3,600 feet.

This is a harmless, nonvenomous mimic of the venomous Fer-de-Lance. It has round pupils, instead of the vertical catlike pupils characteristic of the real Fer-de-Lance. Its imitation of a much more venomous snake makes it less likely to be approached by predators or people.

A diurnal snake, this rear-fanged species is a toad specialist. It hunts for *Bufo* toads on the ground. Frogs and tadpoles are also eaten. When a toad is grabbed, its instinctive response is to inflate its body to prevent being swallowed. The False Fer-de-Lance, however, maneuvers the toad so that the fangs at the rear of its mouth puncture the toad's skin and deflate the toad so that it can be swallowed. Preferred habitats include primary and secondary forests in lowland and premontane elevations on both the Caribbean and southwestern Pacific slopes.

Although not venomous, this snake puts on an impressive display of aggressive behavior if approached. It raises its head, flares the skin at the sides of the head like a viper, hisses, and strikes. The snake shown here was encountered during a daytime walk on the forest trail at Rancho Naturalista.

False Fer-de-Lance, a nonvenomous snake

CORAL SNAKE AND FALSE CORAL SNAKE

Coral snake, showing yellow bands adjacent to red bands on this venomous snake

The brightly colored coral snake is widely distributed but seldom seen. It has alternating red, yellow, and black rings around the body. Although this snake is venomous, it is nocturnal and burrows under leaf litter, so encounters with humans are infrequent. The bright colors provide a warning that deters large predators. There are four species of coral snakes in Costa Rica: *Micrurus nigrocinctus* (shown here), *M. mipartitus* (bicolored with black-white and black-pink bands), *M. clarki,* and *M. alleni.* Each has a different arrangement of colored rings along the body. The False Coral Snake (*Erythrolamprus bizona,* family Colubridae) shown below looks similar to a coral snake, but it is not venomous. The best feature to use to distinguish between the two snakes is that the False Coral Snake has black bands adjacent to red bands and the coral snake has red bands adjacent to yellow bands. The phrase "Red on black, friend of Jack; red on yellow, kill that fellow," is used to help remember the difference. Just because coral snakes are venomous does not mean that they should be killed. Most coral snake bites occur when people try to pick up and handle the snake—not a smart thing to do.

Daytime hours are spent concealed in leaf litter, hollow logs, or ant hills. If approached, a coral snake warns of its presence by raising its coiled tail and waving it back and forth; it also swings the head from side to side with the mouth open.

This snake occupies dry and riparian forests in Guanacaste, moist and wet forests of the Caribbean and southern Pacific lowlands, and forests of middle elevations. It feeds by crawling under leaf litter and biting small lizards, snakes, and amphibians. The prey are immobilized by nerve toxin in the venom and then swallowed. This venom can be fatal to humans, but the snake's mouth is so small that it is difficult for the snake to bite a human. Again, most biting incidents occur when careless people try to handle these snakes.

Micrurus nigrocinctus

Costa Rican names: *Coral; coralillo.*

4/23 trips; 6 sightings.

Total length: 24–36 inches (61.0–91.4 centimeters).

Range: Southern Mexico to northwestern Colombia.

Elevational range: Sea level to 4,000 feet.

False Coral Snake, showing black bands adjacent to red bands on this nonvenomous snake

EYELASH VIPER (PALM VIPER)

Bothriechis schlegelii (formerly Bothrops schlegelii)

Costa Rican names: *Bocaracá*; *oropel* (golden morph).

2/23 trips; 3 sightings.

Total length: 19.7 inches (50 centimeters).

Range: Southern Mexico to Ecuador and Venezuela.

Elevational range: Sea level to 1,700 feet.

The Eyelash Viper is a small but very venomous snake. It inhabits Costa Rica's Caribbean and southern Pacific lowlands and foothills and the forests of the Santa Elena region at Monteverde. The name "eyelash" refers to the hoodlike scales over each eye. This arboreal species spends much of its life resting motionless on palm tree trunks, branches, and fruits awaiting its prey, which explains its other common name. This amazing snake comes in six colors: green, brown, rust, gray, light blue, and gold. The golden morph, known only from Costa Rica, blends into its surroundings when concealed on ripe yellow palm fruits.

Prey includes small lizards, frogs, hummingbirds, and small rodents. Because this snake is so well camouflaged, it is difficult to spot. Bites can occur when a person tries to climb trees or vines or reaches into tree branches or clusters of tropical fruits without inspecting them first. Several fatalities occur in Costa Rica each year as a result of bites from this snake.

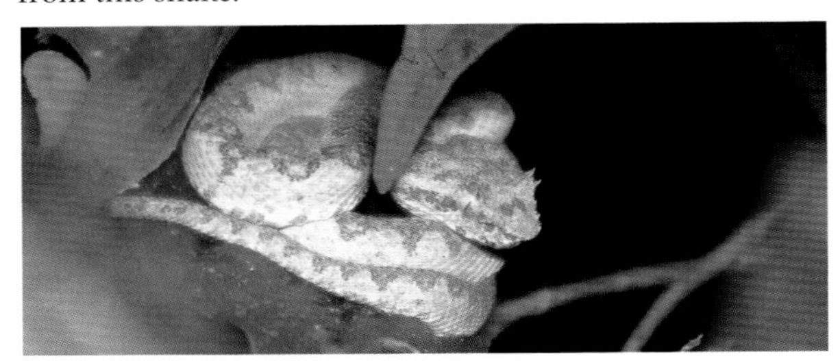

Eyelash Viper, pale gray color phase

Eyelash Viper, gray color phase

Eyelash Viper, golden color phase

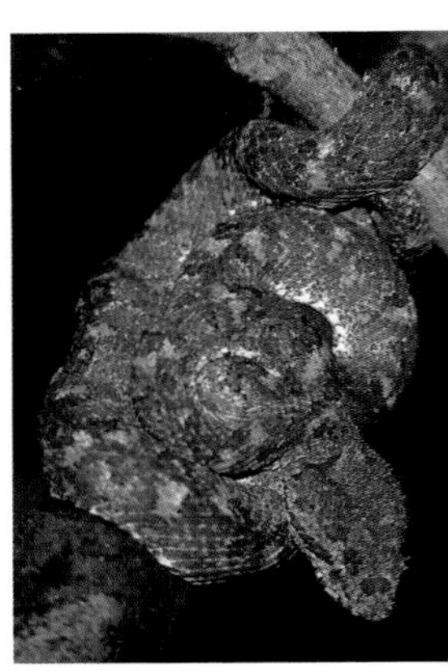

Eyelash Viper, green color phase

FER-DE-LANCE

Fer-de-Lance adult, also known as *terciopelo* in Costa Rica

The legendary Fer-de-Lance is one of the most famous venomous snakes in Costa Rica. An adult can attain a length over six feet. A member of the pit viper family, it has the large, triangular head typical of pit vipers. Pit vipers are named for their large and conspicuous heat-sensitive pits between the nostrils and the eyes.
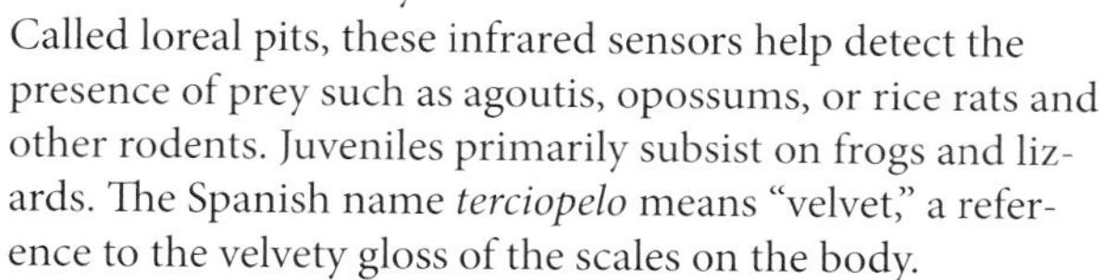
Called loreal pits, these infrared sensors help detect the presence of prey such as agoutis, opossums, or rice rats and other rodents. Juveniles primarily subsist on frogs and lizards. The Spanish name *terciopelo* means "velvet," a reference to the velvety gloss of the scales on the body.

Bothrops asper

Costa Rican name: *Terciopelo.*

1/23 trips; 1 sighting.

Total length: Up to 98.4 inches (250 centimeters).

Range: Northern Mexico to Ecuador and Colombia.

Elevational range: Sea level to 4,250 feet.

This is the most dangerous venomous snake in Central America. When it bites, its long fangs usually deliver a large dose of hemotoxic venom. It is aggressive, and its cryptic markings make it hard to spot as it rests quietly waiting for prey to pass along a trail. According to Savage (2002), about 250 bites of the Fer-de-Lance are recorded in Costa Rica annually, and about fifteen to twenty of those bites result in death. People come into contact with the Fer-de-Lance mainly on banana plantations, where the snake is attracted by high numbers of rats. As a result, the likelihood of encountering this snake is higher for field workers than it is for tourists.

The Fer-de-Lance remains concealed in cover during the day; at night it comes out to wait beside a trail for passing mammals. This species can be found in open, grassy cover, near human settlements, and in deep forests.

The mating season occurs from September to November in Pacific moist and wet forest lowlands and from February to April in the Caribbean moist and wet forest lowlands. After a gestation period of six to eight months, females give birth to a live litter of five to eighty-six young. The young males have a yellow tip on the tail; it may be an adaptation for attracting frogs and lizards when the snake vibrates its tail. The yellow tail is the source of the Spanish nickname *rabo amarillo.*

Young Fer-de-Lance

Preferred habitats include both primary and secondary forests and plantations at lowland and premontane forest elevations. If someone approaches too closely, this snake will vibrate its tail and create a buzzing sound to warn of its presence.

In spite of this snake's ominous reputation, there is little likelihood that ecotourists will encounter any venomous snakes if they are accompanied by a professional naturalist guide who keeps them on main roads and open forest trails, where the chance of finding snakes is low and the chance of spotting snakes at a distance is better. During twenty-three birding trips to Costa Rica since 1987, the equivalent of nearly one year in the field in the national parks, wildlife refuges, and ecotourism resorts of Costa Rica, the author has encountered only one small Fer-de-Lance. In other words, venomous snakes are very hard to find in Costa Rica, and they are no more interested in finding you than you are in finding them.

A Fer-de-Lance awaiting prey. Note how well it is camouflaged.

CENTRAL AMERICAN BUSHMASTER

Lachesis stenophrys

Costa Rican name: *Matabuey; cascabela muda.*

0/23 trips; 0 sightings.

Total length: Up to 153 inches (390 centimeters).

Range: Nicaragua to Panama.

Elevational range: Sea level to 2,000 feet.

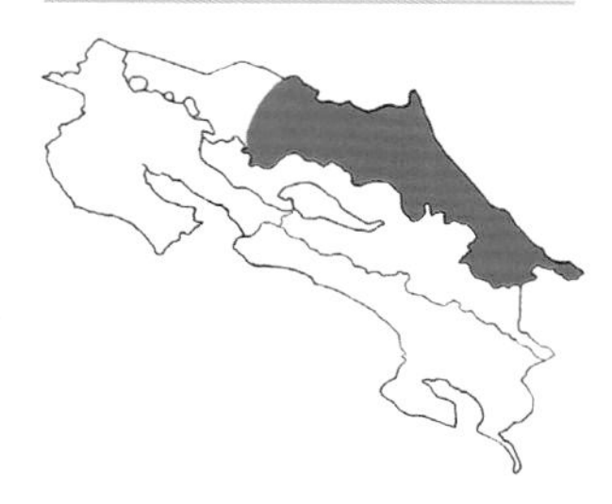

There are two species of bushmasters in Costa Rica. The Central American Bushmaster (*Lachesis stenophrys*) lives in the Caribbean lowlands and is the largest pit viper in the Americas (specimens may exceed seven feet in length). The Spanish name *matabuey* means "ox-killer." The second species of bushmaster, found in extreme southwestern Costa Rica and adjacent areas of Panama, is the Black-headed Bushmaster (*Lachesis melanocephala*). It is similar in behavior and habitat preferences to the Central American Bushmaster.

In contrast to the Fer-de-Lance, which is an aggressive species, the bushmasters are shy and retiring. They inhabit primary rainforests of moist and wet forest lowlands. Because this species is uncommon and is confined to rainforest habitats, only two to four bites are reported per year in Costa Rica. The large size of this snake and the large amount of venom it delivers causes about 75 percent of those bites to be fatal. When approached, this snake may vibrate its tail loudly in the ground litter to warn of its presence.

Among the main prey of this snake is the spiny rat (*Proechimys*), which feeds on the fruits of the *Welfia* palm in the rainforest understory. A bushmaster will sometimes wait for rats in the vicinity of these palms. Bushmasters are the only pit vipers that lay eggs. The female lays ten to twelve eggs that require about sixty days for incubation.

This species is extremely rare; the author has encountered it only once in Costa Rica, in 1969. The snake had just been killed by a laborer who was clearing a trail at La Selva Biological Field Station. It was made available for photography before being prepared as a specimen for the University of Costa Rica by herpetology professor Douglas Robinson. In other words, don't worry about encountering a bushmaster in Costa Rica. It probably won't happen.

Bushmaster encountered by the author in Costa Rica in 1969

Bushmaster adult

TROPICAL RATTLESNAKE

Crotalus durissus

Costa Rican name: *Cascabel.*

0/23 trips; 0 sightings.

Total length: Up to 70.9 inches (180 centimeters).

Range: Mexico to Argentina.

Elevational range: Sea level to 5,400 feet.

The Tropical Rattlesnake, the only rattlesnake in Costa Rica, is found primarily in dry forest habitats of the Guanacaste region. It is characterized by dark diamond-shaped marks on the back and a variable number of buttons composing the rattle on its tail.

Although this rattler is large and has a reputation as an aggressive and venomous snake, only about five bites are recorded per year in Costa Rica, and few are fatal. It is nocturnal, and if disturbed during the day it will typically raise its head, hiss, and vibrate its rattle as a warning.

Prey includes small mammals and ctenosaurs, which are abundant in Guanacaste. Mating occurs in December and January. Females give birth to live young about six months later. Litters vary from fourteen to thirty-five young. Preferred habitat includes grasslands, savanna areas, and dry forest habitats throughout Guanacaste. During twenty-three birding trips to Costa Rica since 1987 (the equivalent of nearly one year of field activities), no rattlesnakes have been encountered. By staying on open main trails and roads during field trips, tour groups minimize the likelihood of encountering venomous snakes. The rattlesnake portrayed here was photographed in captivity.

Tropical Rattlesnake photographed in captivity

WHITE-LIPPED MUD TURTLE

Kinosternon leucostomum

Costa Rican names: *Tortuga caja; tortuga amarilla; candado pequeño.*

3/23 trips; 3 sightings.

Total carapace length: 7 inches (17.8 centimeters).

Range: Veracruz, Mexico, to Colombia, Ecuador, and Peru.

Elevational range: Sea level to 4,000 feet.

Among the more interesting freshwater turtles in Costa Rica is a mud turtle that has two hinges on its plastron, allowing the turtle to close its shell completely when approached. The high-domed shell has a single keel along the midline of the carapace and an elongated oval profile when viewed from above. Both features are characteristic of this species. The overall body color is dark brown to blackish on the carapace, and the plastron is dark yellow. The jaws and chin are yellowish. Males have a relatively long tail, and females have a very short tail.

Nesting occurs twice each year, with major nesting activity in July and October. Multiple clutches of one to five eggs (usually one) are laid on the ground and covered with leaf litter. The eggs hatch in 126–148 days. This turtle is considered nocturnal, but it can occasionally be found crossing roads between wetlands during the day. The White-lipped Mud Turtle eats aquatic plants, mollusks, insects, worms, carrion, and aquatic invertebrates.

This turtle is found in quiet waters of marshes, swamps, ponds, rivers, and streams in lowlands and middle elevations of the Caribbean and southern Pacific slopes. It can also be seen on adjacent upland sites. Look for it in the wetlands of Carara, Corcovado, Cahuita, Manuel Antonio, and Tortuguero NPs; La Selva Biological Field Station; Caño Negro NWR; and near San Vito.

White-lipped Mud Turtle

White-lipped Mud Turtle ventral surface, with two hinges that allow it to close

GREEN TURTLE

Chelonia mydas

Costa Rican names: *Tortuga verde; tortuga blanca.*

Total length: Average 39.3 inches, with records to 60.2 inches (152.9 centimeters).

Weight: 130–440 pounds (59–200 kilograms); record is 850 pounds (386.4 kilograms).

Range: Atlantic, Pacific, and Indian Oceans.

The Green Turtle is one of four hard-shelled sea turtles found along Costa Rica's coasts. Others include the Hawksbill, Olive Ridley, and Loggerhead. Although the Green Turtle nests at other beaches in the Caribbean, the Tortuguero site is one of the most significant in the world. It hosts 5,000–15,000 turtles along its twenty-two miles of beach from June through November, with a peak in late August.

Females may reproduce only once every two to three years. After mating offshore, females go ashore to nest three to seven times at thirteen-day intervals. The peak of nesting is at night during high tide. About 100–150 eggs are laid each time. The eggs hatch after incubating in the sand for forty-five to sixty days. The sex of the baby turtles is determined by temperature. Warmer temperatures create more females, and cooler temperatures create more males.

Green Turtles mainly eat plants. Extensive turtle-grass beds off the Mosquito Coast of Nicaragua are a primary feeding area. They also eat mangrove roots and leaves and green, brown, and red algae. Occasional animal foods include small mollusks, crustaceans, sponges, and jellyfish. These turtles may live thirty to fifty years, but they

Green Turtle, one of Costa Rica's most famous endangered species

have become endangered because they are killed for their meat. Their eggs are also collected for use as food and as an aphrodisiac.

Costa Rica has a worldwide reputation as a major Green Turtle nesting area because of the pioneering work started at Tortuguero in 1954 by the late Dr. Archie Carr. His book *So Excellent a Fishe* documents his conservation story. As many as 15,000 people now come to Tortuguero each year to see nesting Green Turtles. Contact the Caribbean Conservation Corporation at 1-800-678-7853 for details on helping turtles and on how to see them there. (Remember that Green Turtles are not nesting during the main tourist season from January through March.)

The Green Turtles in the Caribbean are mainly brown and do not have indentations on the carapace above the hind legs. Green Turtles are also found along the Pacific coast of tropical America and are sometimes regarded as a separate species, *Chelonia agassizii*. They are more greenish to olive-brown and have conspicuous indentations on the carapace above the hind legs. Most of these turtles are very dark brown or black on the upper portions of the soft parts. The head scales are light-margined in *C. mydas* and uniformly dark on *C. agassizii*. In the Galápagos Islands, these are referred to as Black Turtles.

The Green Turtle Research Station at Tortuguero founded by Dr. Archie Carr

The famous nesting beach used by Green Turtles at Tortuguero

OLIVE RIDLEY SEA TURTLE (PACIFIC RIDLEY)

Lepidochelys olivacea

Costa Rican names: *Lora* (Playa Ostional); *carpintera* (Playa Nancite).

Total carapace length: 25.1–28 inches (63.8–71.1 centimeters).

Weight: 88 pounds 3 ounces (40 kilograms).

Range: Atlantic, Indian, and Pacific Oceans in tropical regions.

Olive Ridley Sea Turtles, the most abundant of all Costa Rican sea turtles, are well known for their *arribadas* in the Guanacaste region at the Nancite and Ostional beaches on the Pacific coast. During earlier *arribadas,* as many as 120,000 Ridley turtles emerged to nest during periods of four to eight days from July through December. In recent years, however, the numbers have declined. The *arribadas* occur at two- to four-week intervals. Smaller *arribadas* occur at these beaches from January through June. A few Olive Ridley Sea Turtles nest along the Pacific coast south to Panama, and tracks leading from their nests may be seen regularly on the beaches from the Sirena Biological Station to Carate on the Osa Peninsula during January and February.

When a female Ridley comes ashore, she digs a hole fifteen to twenty inches deep with her hind flippers. About one hundred eggs are deposited and covered with sand using the hind flippers. A female normally nests two times in one season at an interval of twenty-eight to thirty days. The incubation period is about fifty days. Hatchling turtles have many predators: frigatebirds, coatis, coyotes, feral dogs, pigs, raccoons, opossums, vultures, sharks, and even crabs. Few hatchlings (probably less than 1 percent) survive their first year.

At Playa Ostional, 20 million to 30 million eggs may be laid during one season. Eggs laid during the first days of an *arribada* are dug up by turtles nesting later. Since many of these eggs go to waste, the villagers of Ostional are organized

An Olive Ridley Sea Turtle crawls ashore at the Ostional Beach on the Pacific Coast. Photo by Pablo Vásquez Badilla.

into the Ostional Development Corporation and are authorized by the government to collect and sell eggs laid at the beginning of an *arribada*. They collect a quota of 3 million eggs per year and receive about $95,000 in income. Some of the money is used to staff the Douglas Robinson Marine Turtle Research Station, and the remainder has been used to build a new school and a health clinic. Turtle eggs from Ostional are sold in small bags that are labeled to show their legal origin. Otherwise, it has been illegal to take or sell sea turtle eggs in Costa Rica since 1966.

Ridley turtles are carnivores and dive as deeply as 500 feet to eat shrimp, crabs, snails, sea urchins, jellyfish, and fish eggs. Studies of turtle migration show that Ridleys seasonally move south to feeding areas offshore from Ecuador. Although the Olive Ridley is protected in Costa Rica, fishermen off the coast of Ecuador kill this turtle in large numbers. Because of the difficulty of predicting when an *arribada* will occur and the remoteness of Playas Nancite and Ostional, it is difficult to experience the wonder of an *arribada*. During the main nesting period, from July through December, a few turtles can be seen nesting every night. Nesting appears to peak during the waning quarter moon of months in the nesting season, especially in July. Visits to Nancite must be coordinated with Santa Rosa NP service officials, and visits to Ostional must be coordinated with Ostional turtle research station biologists.

Ridley turtle hatchlings, near Playa Grande

Olive Ridley Sea Turtles lay their eggs at Ostional. This mass nesting phenomenon is called an *arribada*. Photo by Pablo Vásquez Badilla.

LEATHERBACK TURTLE

Dermochelys coriacea

Costa Rican name: *Baula.*

4/23 trips; 4 sightings.

Total length: 70.8 inches (179.8 centimeters)

Weight: Up to 1,300 pounds (591 kilograms); one record of 2,000 pounds (909 kilograms).

Range: Atlantic, Pacific, and Indian Oceans, including tropical and temperate waters.

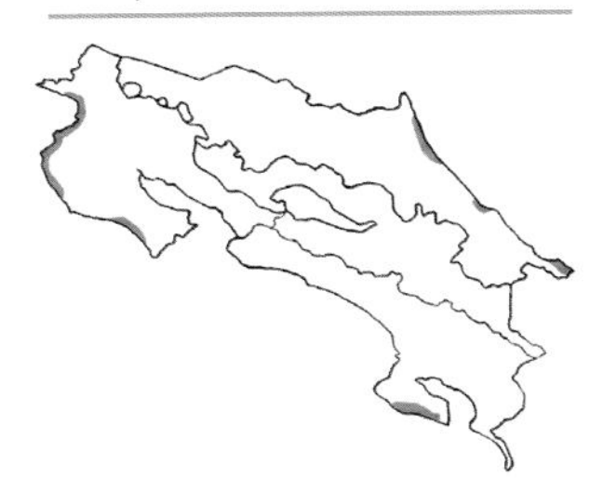

Leatherback Turtle adult

The Leatherback Turtle is the largest reptile in the world. A male accidentally caught off the coast of Wales measured nine feet long and weighed 2,000 pounds. In contrast to most other sea turtles, which have hard shells, the Leatherback's shell has a firm, leathery texture. The upper shell (carapace) has seven parallel dorsal ridges and averages sixty-three inches long and forty-four inches wide.

The flexible shell is an adaptation for diving as deep as 3,330 feet, where the pressure exceeds 1,500 pounds per square inch. Another diving adaptation is the storage of oxygen in muscle tissues when they dive, because the pressure would crush their lungs. They dive to these depths in search of jellyfish. Other foods include sea urchins, squid, crustaceans, mollusks, fish, blue-green algae, and seaweeds. Sometimes they die from eating discarded plastic bags and balloons, which to them resemble jellyfish.

Playa Grande near Tamarindo on the Pacific coast is one of thirteen primary Leatherback nesting beaches in the world. Up to 1,600 Leatherbacks have nested there in one season, which extends from October through February. Individual females come ashore repeatedly at ten-day intervals, and each female may nest three to six times during the nesting season. Nesting typically occurs at night during high tide. About one hundred eggs are deposited in a nest hole that is dug with the hind flippers. The eggs hatch after sixty days. The number of nesting turtles declined to only thirty-two in 2009.

This endangered species faces threats from ocean garbage, egg poaching, predation on hatchlings by dogs and pigs, illegal killing for meat by commercial fishermen, and

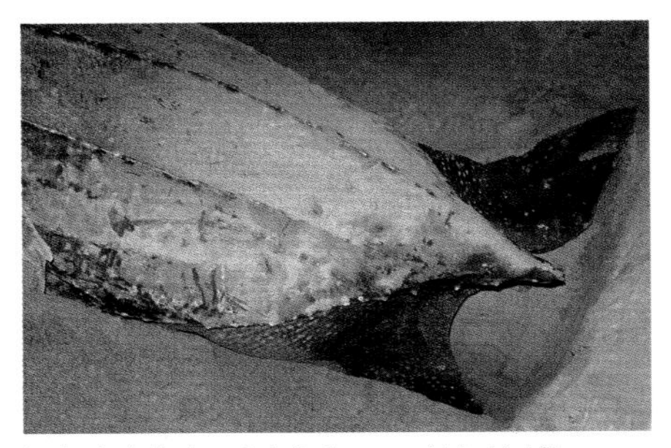

Leatherback digging a hole for its eggs with its hind flippers

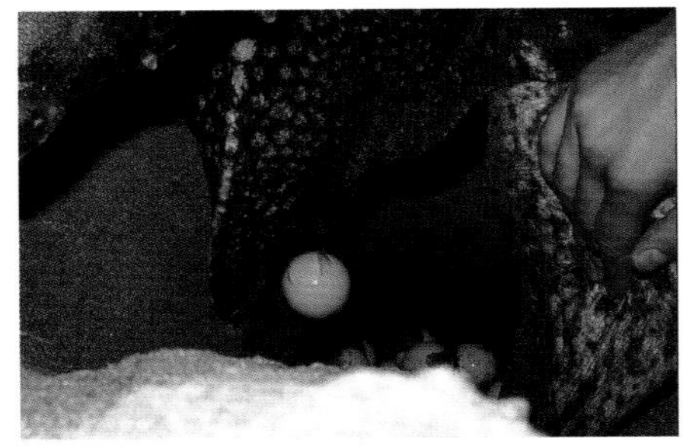

Leatherback in the process of laying an egg

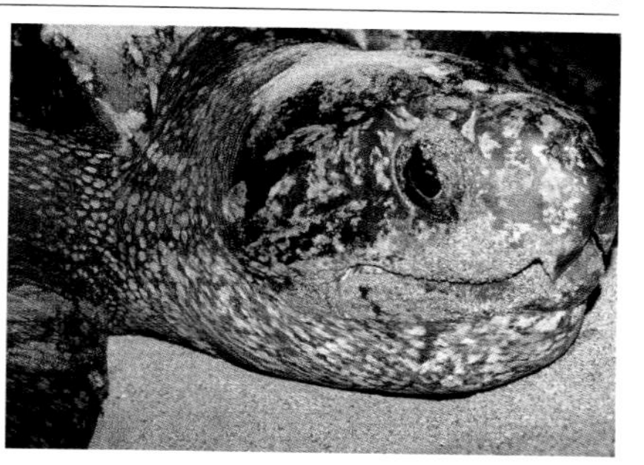

Leatherback Turtle female

lights on beaches that disorient the young after hatching. The world population was estimated to have declined from 115,000 to 34,500 since 1980, but the largest nesting concentration of Leatherbacks in the world has been discovered along the coast of Gabon in West Africa. The population there may include 16,000 to 41,000 additional turtles. The half-mile-long nesting beach at Playa Grande was designated as Las Baulas NP in 1995. A few Leatherbacks also nest on the beaches at Tortuguero NP, along the coast of the Osa Peninsula, and at Corcovado Lodge Tent Camp.

Since Leatherbacks nest during the peak of Costa Rica's tourist season, this is the species most easily seen in January and February. It is worth the pilgrimage to Tamarindo to see the Leatherbacks. Sea turtles have been on this earth for at least 150 million years. That is why sitting on a Costa Rican beach at night watching a sea turtle plod ashore, dig its nest, and lay its eggs becomes a mystical, almost religious, experience. It transports you back in time millions of years; it connects you with ancient rhythms and natural processes that help you understand and appreciate the incredible diversity of life that is preserved in Costa Rica's national park system. Check with nature tourism companies in Costa Rica or with hotels in the Tamarindo vicinity to arrange for turtle-watching tours at Playa Grande. At Playa Grande you must be escorted by guides from the National Park Service to see the turtles. Las Baulas NP is currently threatened by developers who seek to privatize land in the park. Interested persons should go to the Leatherback Trust Web site, www.leatherback.org, for more information.

Leatherback Turtle hatchling

Typical Leatherback nest, with about 100 eggs

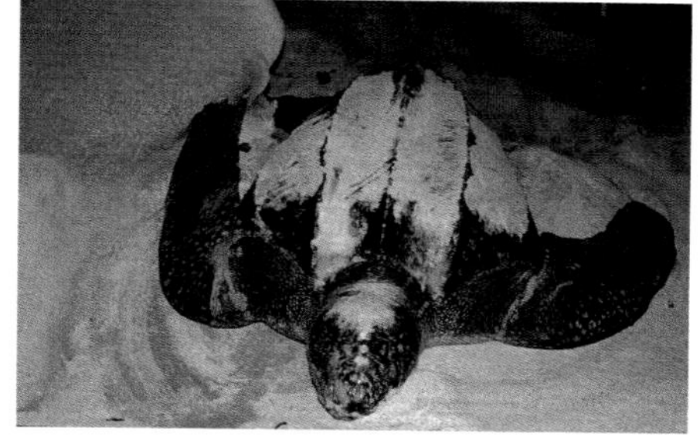

After laying the clutch of eggs, the turtle fills the nest hole with great strokes of its flippers.

BLACK RIVER TURTLE (BLACK WOOD TURTLE)

Rhinoclemmys funerea

Costa Rican names: *Tortuga negra del río; jicote.*

11/23 trips; 17 sightings.

Total carapace length: 12.8–14.0 inches (32.5–35.6 centimeters).

Weight: 10 pounds (4.5 kilograms).

Range: Caribbean lowlands on the border of Honduras and Nicaragua to Panama.

Elevational range: Sea level to 3,240 feet.

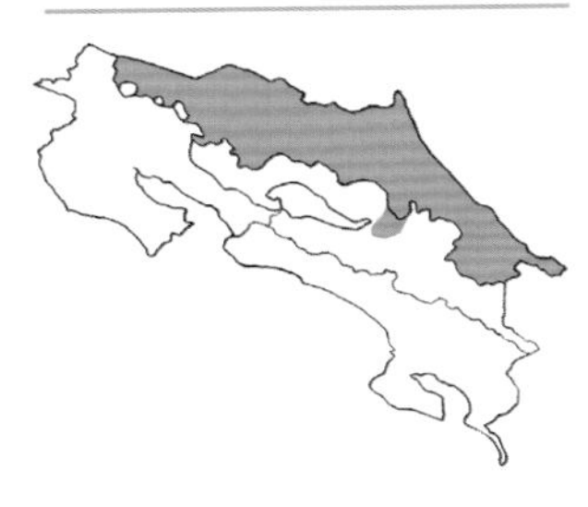

The Black River Turtle is the largest and most conspicuous turtle in rivers of the Caribbean lowlands and middle elevations. This aquatic turtle frequents ponds, rivers, and swamps, where it can typically be seen basking on partially submerged logs. The Black River Turtle has a high-domed shell that is dark brown to black. The head, neck, and edges of the shell have yellow highlights.

This turtle eats fruits, grasses, and broad-leaved plants. The nesting season extends from March through August. A female lays eggs one to four times per season, with an average clutch of three eggs. The incubation period is 98–104 days. Eggs are laid on the ground and covered with leaves. This turtle is commonly seen during the day along the canals of Tortuguero NP or the Río San Juan and its tributaries, from the Stone Bridge at La Selva Biological Field Station, and along the boat canal from Limón to Tortuguero.

Black River Turtle

ORANGE-EARED SLIDER

Slider turtles of the species *Chrysemys ornata* include fourteen subspecies that range from Virginia to Brazil. In northern regions, it is called the Red-eared Slider because of the red mark above and behind the eyes. The two Costa Rican subspecies are very similar and have orange markings. They can be seen basking on floating logs in freshwater wetlands and along lowland rivers of the Caribbean and Pacific slopes. This turtle seems to prefer wetlands with muddy or murky bottoms. Sometimes they can be seen stacked several turtles deep on their favorite basking logs.

Young turtles eat tadpoles, small fish, crayfish, shrimp, and snails. As they mature, their diet broadens to include herbaceous plants and algae. The nesting season extends from December to May. A female will lay several clutches of two to thirty-five eggs during the dry season. The young hatch between 69 and 123 days. This turtle should be looked for along the canals of Tortuguero NP and in nearby woodland ponds and backwaters like those at Tortuga Lodge.

Chrysemys ornata (formerly Trachemys scripta venusta and T. s. emolli)

Costa Rican name: *Tortuga resbaladora.*

2/23 trips; 2 sightings.

Total carapace length: Up to 15 inches (38.1 centimeters).

Range: Mexico to northern Argentina.

Elevational range: Sea level to 3,000 feet.

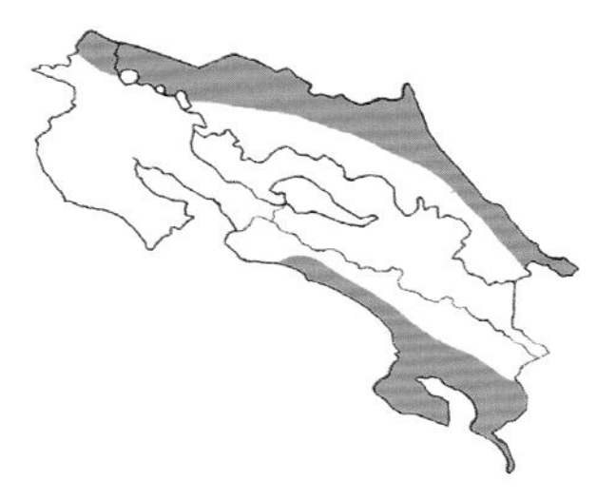

Orange-eared Slider

SPECTACLED CAIMAN

Caiman crocodilus
Costa Rican names: *Caimán; lagarto; baba; babilla.*
7/23 trips; 11 sightings.
Total length: 4–6 feet (120–183 centimeters).
Range: Southern Mexico to Ecuador and central Brazil.
Elevational range: Sea level to 1,000 feet.

The Spectacled Caiman is related to alligators, but it is smaller than the American Alligator and does not pose a threat to humans. The caiman occupies lowland wetlands, rivers, marshes, ponds, and even water-filled roadside ditches of the Caribbean and Pacific slopes. Prey consists of fish and amphibians, though other kinds of wildlife may be eaten as carrion. Newly hatched caimans eat insects. Young caimans are eaten by Jabirus, Wood Storks, Great Egrets, and raccoons. Adult caimans have no predators except human poachers.

Nesting occurs mainly during February and March. Females lay and bury twenty to forty eggs in a nest mound near the water's edge. The eggs hatch after seventy-three to seventy-five days. Upon hatching, the young are eight to nine inches long; they call to the parents, who open the nest and carry the young to the water in their mouth. The young are defended by the parents for at least four months after hatching.

Caimans may be viewed from boats along the canals of Tortuguero NP, along lowland rivers, and on wetlands at La Laguna del Lagarto Lodge. Distribution includes lowlands throughout Costa Rica, but caimans are uncommon in Guanacaste. During the day, caimans lie quietly in beds of water hyacinth so that only their eyes and nostrils are exposed. At night, when spotlights are shined on them, the eyes reflect light like red coals along the edges of rivers and ponds.

Spectacled Caiman, basking in the sun

Spectacled Caiman

AMERICAN CROCODILE

Crocodylus acutus

Costa Rican name: *Cocodrilo.*

18/23 trips; 45 sightings.

Total length: Usually 10–12 feet (300–370 centimeters); record is 23 feet (700 centimeters).

Range: Florida to Ecuador.

Elevational range: Sea level to 650 feet.

The American Crocodile adds an imposing and prehistoric presence to Costa Rica's rivers, swamps, and wetlands. Although the caiman seldom exceeds six feet, the crocodile is frequently twice that length. A crocodile can be distinguished from a caiman by the shape and structure of the snout. The snout is more pointed on a crocodile, and the canine teeth of the lower jaw are visible because they fit in grooves on the sides of the upper jaw. The snout of the caiman is more broadly rounded, and the canines of the lower jaw are not visible because they fit into pits in the upper jaw.

Like caimans, the American Crocodile builds a nest mound near the water's edge that contains about twenty eggs. Upon hatching, the young call to the female, who digs them out and gently transports them to the water in her mouth. The young are protected by the adult for an extended period after hatching.

This is an endangered species because it has been relentlessly killed in the past. Protection has allowed the crocodile to make an impressive recovery. It occurs in salt, brackish, and freshwater habitats. The diet includes fishes, turtles, and mammals up to the size of deer. When a crocodile swallows its prey, it must tip its head upward because it cannot swallow with its head in a horizontal position. Since the late 1990s, crocodiles have killed at least four people in Costa Rica; the casualties were swimming in rivers inhabited by crocodiles.

American Crocodile, profile

Sunlight helps kill bacteria in the mouth of the basking crocodile.

Among the places where crocodiles can be seen are the canals in Tortuguero NP, along the Río Colorado and Río San Juan, the marshes of Palo Verde NP, the Río Tempisque and its tributaries in Guanacaste, the Río General southeast of San Isidro del General, and even such unlikely places as the city water treatment lagoons at San Isidro del General. There are four excellent places to observe crocodiles safely. The first is from the bridge over the Río Grande de Tárcoles near the entrance to Carara NP on the Pacific coast. More than a dozen crocodiles can routinely be observed sunning themselves on sand bars near the bridge. The other locations are at Hacienda Solimar, in the lagoon at Estero Madrigal surrounding the heronry; in the mangrove lagoons and wetlands at La Ensenada Lodge; and during boat tours in the mangrove lagoons near Quepos.

A crocodile quietly awaits its prey by looking like a log in the water.

A crocodile enters the Río San Juan along the Nicaragua border.

GLOSSARY

Amplexus: The position in which a male frog holds onto a female frog during mating. The male grasps the female from above, around her chest or groin area, with his forelegs.

Anticoagulant: A chemical compound that prevents blood from clotting. Some snakes have venom that functions as an anticoagulant.

Aquatic: Associated with water, typically fresh water in a pond, lake, marsh, stream, or river, but also a microhabitat in a bromeliad or other water container.

Arboreal: Adapted to living in trees (e.g., squirrels and monkeys).

Arribada: A massive, synchronized arrival of sea turtles onto a beach for nesting purposes.

Arthropod: A member of the phylum Arthropoda, including segmented invertebrates like insects, spiders, and crustaceans.

Bask: To rest in the sun, as commonly practiced by turtles, caimans, and other reptiles to raise their body temperature.

BR: Biological Reserve.

Canopy: A layer within the overall structure of a forest. *Lower canopy* refers to shrub layer and short trees; *midcanopy* refers to middle levels of foliage and structure among trees; *upper canopy* refers to the foliage and vegetative structure in the tops of the trees; and *supercanopy* refers to trees like *Ceiba* that project above the upper canopy.

Carapace: The upper shell of a turtle.

Caribbean slope: The northern and eastern portion of Costa Rica from the continental divide in the mountains eastward to the lowlands that drains toward the Caribbean.

Clutch: The eggs laid in a single bird's or reptile's nest, usually laid by a single female. (After hatching, the young are collectively referred to as a brood.)

Color morph or phase: A color variety within a species that deviates from the normal coloration. For example, the Jaguar has a black color morph, and there are dark or black morphs among some birds of prey.

Crepuscular: Being active at dawn and dusk.

Crustacean: A type of arthropod that typically has a hard shell covering the body; includes lobsters, shrimp, and crabs.

Cryptic: Difficult to see because of markings or colors that blend in with the background.

Dewlap: A fold of fleshy skin that hangs down from the chin and throat area, as on an iguana or anole.

Diurnal: Being active during daylight hours; opposite of nocturnal.

Dry forest: A forest in which the range of total annual rainfall is 40–80 inches per year. The dry forest in Costa Rica is found primarily in Guanacaste Province and is a tropical deciduous forest with distinct wet and dry seasons. The dry season occurs from December to March.

Dry season: That portion of the year in which less rainfall occurs. In Costa Rica, the dry season occurs from December through March and is most prevalent on the Pacific slope.

Endemic: Occurring only within a limited geographic area and nowhere else in the world, such as species found only in the mountains of Costa Rica and western Panama, only in the lowlands of southeastern Costa Rica and adjacent Panama, or only on Cocos Island.

Epiphyte: A plant that grows above the ground on the surface of another plant and depends on that plant for physical support. Examples include some bromeliads and orchids.

Foothills: Hilly terrain at the base of mountains with elevations of about 1,500–4,500 feet (approximately the same as the premontane or subtropical zone). The distribution of some species is limited to foothills and not to either lowlands or highlands.

Gallery forest: A forest adjacent to a stream or river; also referred to as riparian forest.

Gestation period: The time from breeding to the birth of young; usually associated with mammals.

Harem: A group of females attended by one male; for example, a male Sac-winged Bat will have a harem of five or six females.

Herpetofauna: Reptiles and amphibians.

Highlands: For the purposes of this book, regions at 5,000–11,000 feet in elevation, including moist, wet, and rainforests. They are characterized by 40–200 inches of rainfall per year. Such areas can have large numbers of epiphytes on the trees, and the area's diversity is lower than at lower elevations. It includes 15 percent of Costa Rica's land area.

Incubation period: For reptiles, the time between egg laying and egg hatching.

Lowlands: Relatively flat terrain between foothills and the coast, in the elevation range from sea level to about 2,000 feet. It includes dry, moist, and wet forests and accounts for 57 percent of Costa Rica's land area.

Melanistic: Having excess pigmentation, as in a color phase in which an individual creature is very dark or black.

Middle elevation: Elevations known as subtropical or premontane, including elevations of about 2,000–4,500 feet; also referred to as foothills. This may include moist, wet, and rainforest habitats. The Central Plateau and San José are at middle elevations. This zone includes 28 percent of Costa Rica's land area.

Mimicry: The phenomenon by which one creature derives a survival advantage from its resemblance to another creature that has special defenses against predation. There are two main kinds of mimicry, Batesian and Müllerian.

Moist forest: A forest in tropical lowlands that receives 80–160 inches of rainfall per year. This type of forest covers 24 percent of Costa Rica's land area. At premontane and lower montane levels, moist forest receives 40–80 inches of rainfall and includes about 5 percent of Costa Rica's land area.

Mollusk: An invertebrate animal, such as a snail, clam, squid, or octopus, with one or two shells protecting all or part of the soft body.

Monogamy: A reproductive strategy by which a single male and female create a pair bond with each other and do not have multiple mates.

Montane forest: A highland wet forest (40–80 inches of rainfall per year) or rainforest (80–160 inches of rainfall per year) at 8,200–10,500 feet in elevation.

Morph: *See* Color morph or phase.

Nocturnal: Active at night; the opposite of diurnal.

NP: National Park.

NWR: National Wildlife Refuge.

Pacific slope: That portion of the Costa Rican landscape that drains from the mountains west and south to the Pacific Ocean.

Paramo: A highland elevational zone above the montane zone, ranging from about 10,500 feet to the peaks of Costa Rica's mountains. The zone is above the tree line and is characterized by stunted shrubs, bamboo, and many composites. It includes 0.2 percent of the country's land area.

Parotoid gland: The large pair of glands found at the back of the head of some toads, containing toxic chemicals that deter predation.

Phase: *See* Color morph or phase.

Plastron: The lower shell of a turtle.

Precocial: Active from birth and needing little parental care; for example, some newly hatched birds are so well developed that they can leave the nest within a day or two of hatching.

Prehensile tail: A muscular tail that is used to curl around branches as an animal feeds in the trees. Examples of mammals with prehensile tails include Tamanduas, spider monkeys, howler monkeys, and White-throated Capuchins.

Premontane: The elevational zone between approximately 2,000 and 4,900 feet; also referred to as foothills, subtropical, or middle elevations. The zone includes moist forests, wet forests, and rainforests at those elevations.

Primary forest: A mature forest that has not been cut in recent times.

Rainforest: A forest that receives more than 320 inches of rainfall per year. The term may also be used more generally to describe moist or wet forests.

Rainy season: That portion of the year in which most rain falls. In Costa Rica, this is the period from April to December.

Riparian forest: A forest along a stream or river; also referred to as gallery forest.

Savanna: An arid habitat of northwestern Costa Rica, near the Nicaragua border, characterized by open grassy ground cover and scattered stunted trees like *Byrsonima* and *Curatella*. It is an arid extreme of tropical dry forest maintained by fire.

Secondary forest: A forest that has grown back after a previous disturbance like burning or cutting.

Subspecies: A distinctive grouping of a population within a species; may be distinguished by its range, coloring, or behavior.

Tadpole: The aquatic larval stage of frogs and toads.

Taxonomist: A person who specializes in the classification and naming of species.

Terrestrial: Living on the ground.

Tico: A person from Costa Rica.

Tympanum: The round area, like an eardrum, immediately behind and slightly below the eye of a frog or lizard.

Understory: Smaller trees, shrubs, and other vegetation that are generally less than twenty-five feet high within a taller forest.

Viviparous: Bearing live young (as opposed to laying eggs).

Wet forest: A tropical forest that receives 160–320 inches of rainfall per year.

Wet season: The season when most rainfall occurs in an area (same as rainy season). In Costa Rica, this is generally April to December. In some areas, such as the northeastern Caribbean lowlands, the wet season extends through most of the year.

BIBLIOGRAPHY

Acuña, Vilma Obando. 2002. *Biodiversidad en Costa Rica*. San José, Costa Rica. Editorial INBIO. 81 pp.

Acuña Mesén, Rafael Arturo. 1998. *Las tortugas continentales de Costa Rica*. 2d ed. San José, Costa Rica: Editorial de la Universidad de Costa Rica. 92 pp.

Andrews, Robin M. 1983. *Norops polylepis*. In *Costa Rican Natural History,* ed. Daniel H. Janzen, 409–410. Chicago: Univ. of Chicago Press. 816 pp.

Banfield, A.W. F. 1974. *The Mammals of Canada*. Toronto, Ont.: Univ. of Toronto Press for the National Museum of Natural Sciences. 438 pp.

Bartlett, R. D., and Patricia Bartlett. 1997. *Lizard Care from A to Z*. Hauppauge, N.Y.: Barron's. 178 pp.

Beletsky, Les. 1998. *Costa Rica: The Ecotravellers' Wildlife Guide*. San Diego, Calif.: Academic Press. 426 pp.

Bergman, Charles. 1999. The Peaceful Primate. *Smithsonian* 30(3): 78–86.

Boinski, Sue, K. Jack, C. Lamarsh, and J. Coltrane. 1988. Squirrel Monkeys in Costa Rica: Drifting to extinction. *Oryx* 32(1): 45–48.

Borrero H., José Ignacio. 1967. *Mamíferos Neotropicales*. Cali, Colombia: Univ. del Valle. 108 pp.

Boza, Mario A. 1987. *Costa Rica National Parks*. San José, Costa Rica: Fundación Neotrópica. 112 pp.

———. 1988. *Costa Rica National Parks*. San José, Costa Rica: Fundación Neotrópica. 272 pp.

Cahn, Robert. 1984. An Interview with Alvaro Ugalde. *The Nature Conservancy News* 34(1): 8–15.

Carr, Archie. 1967. *So Excellent a Fishe: A Natural History of Sea Turtles*. Garden City, N.Y.: Natural History Press. 248 pp.

———. 1983. *Chelonia mydas*. In *Costa Rican Natural History,* ed. Daniel H. Janzen, 390–392. Chicago: Univ. of Chicago Press. 816 pp.

Carr, Archie III. 1998. The Big Green Seafood Machine. *Wildlife Conservation* 101(4): 16–23.

Carr, Archie, and David Carr. 1983. A Tiny Country Does Things Right. *International Wildlife* 13(5): 18–25.

Carrillo, Eduardo, Grace Wong, and Joel C. Sáenz. 1999. *Mamíferos de Costa Rica*. Santo Domingo de Heredia, Costa Rica: Instituto Nacional de Biodiversidad. 250 pp.

Chalker, Mary W. 2007. *Exploring Costa Rica 2008/9*. San José, Costa Rica: Tico Times. 464 pp.

Coborn, John. 1994. *Green Iguanas and Other Iguanids*. Neptune City, N.J.: T.H.F. Publications. 64 pp.

Cornelius, Stephen E. 1983. Olive Ridley Sea Turtle (*Lepidochelys olivacea*). In *Costa Rican Natural History,* ed. Daniel H. Janzen, 402–405. Chicago: Univ. of Chicago Press. 816 pp.

———. 1986. *The Sea Turtles of Santa Rosa NP*. San José, Costa Rica: Fundación de Parques Nacionales. 65 pp.

Crump, Marty L. 1983. Poison Dart Frogs (*Dendrobates granuliferous*) and (*Dendrobates pumilio*). In *Costa Rican Natural History,* ed. Daniel H. Janzen, 396–398. Chicago: Univ. of Chicago Press. 816 pp.

de la Rosa, Carlos L., and Claudia C. Nocke. 2000. *A Guide to the Carnivores of Central America*. Austin: Univ. of Texas Press. 244 pp.

Dixon, Jim R., and Maureen A. Staton. 1983. Spectacled Caiman (*Caiman crocodilus*). In

Costa Rican Natural History, ed. Daniel H. Janzen, 387–388. Chicago: Univ. of Chicago Press. 816 pp.

Echternacht, Sandy C. 1983. Ameiva Lizard (*Ameiva*) and *Cnemidophorus*. In *Costa Rican Natural History*, ed. Daniel H. Janzen, 375–379. Chicago: Univ. of Chicago Press. 816 pp.

Ehrenfeld, David. 1989. Places. *Orion Nature Quarterly* 8(3): 5–7.

Eisenberg, John F. 1989. *Mammals of the Neotropics*. Vol. 1, *The Northern Neotropics*. Chicago: Univ. of Chicago Press. 449 pp.

Emmons, Katherine M., Robert H. Horwich, James Kamstra, Ernesto Saqui, James Beveridge, Timothy McCarthy, Jan Meerman, Scott C. Silver, Ignacio Pop, Fred Koontz, Emiliano Pop, Hermelindo Saqui, Linde Ostro, Pedro Pixabaj, Dorothy Beveridge, and Judy Lumb. 1996. *Cockscomb Basin Wildlife Sanctuary: Its History, Flora, and Fauna for Visitors, Teachers, and Scientists*. Caye Caulker, Belize: Producciones de la Hamaca; and Gays Mills, Wisc.: Orang-utan Press. 334 pp.

Emmons, Louise. 1997. *Neotropical Rainforest Mammals: A Field Guide*. 2d ed. Chicago: Univ. of Chicago Press. 307 pp.

Emmons, Louise, and François Feer. 1997. *Neotropical Rainforest Mammals: A Field Guide*. 2d ed. Chicago: Univ. of Chicago Press. 396 pp.

Emmons, Louise, Bret M. Whitney, and David L. Ross Jr. 1997. *Sounds of Neotropical Rainforest Mammals: An Audio Field Guide*. Chicago: Univ. of Chicago Press, for Cornell Laboratory of Ornithology.

Ernst, Carl H. 1983. Black River Turtle (*Rhinoclemmys funerea*). In *Costa Rican Natural History*, ed. Daniel H. Janzen, 417–418. Chicago: Univ. of Chicago Press. 816 pp.

Ernst, Carl H., and Roger W. Barbour. 1989. *Turtles of the World*. Washington, D.C.: Smithsonian Institution Press. 313 pp.

Fitch, Henry S., and Jenny Hackforth-Jones. 1983. Ctenosaur (*Ctenosaura similis*). In *Costa Rican Natural History*, ed. Daniel H. Janzen, 394–396. Chicago: Univ. of Chicago Press. 816 pp.

Flaschendrager, Axel, and Leo Wijffels. 1996. *Anolis*. Berlin: Terrarien Bibliothek. 207 pp.

Franke, Joseph. 1997. *Costa Rica's National Parks and Preserves: A Visitor's Guide*. Seattle: The Mountaineers. 223 pp.

Glander, Kenneth E. 1996. *The Howling Monkeys of La Pacífica*. Durham, N.C.: Duke Univ. Primate Center. 31 pp.

Gómez, Luis Diego, and Jay M. Savage. 1989. Searchers on That Rich Coast: Costa Rican Field Biology, 1400–1980. In *Costa Rican Natural History*, ed. Daniel H. Janzen, 1–11. Chicago: Univ. of Chicago Press. 816 pp.

Greene, Harry W. 1983. Boa Constrictor (*Boa constrictor*). In *Costa Rican Natural History*, ed. Daniel H. Janzen, 380–382. Chicago: Univ. of Chicago Press. 816 pp.

———. 1983. *Micrurus nigrocinctus*. In *Costa Rican Natural History*, ed. Daniel H. Janzen, 406–408. Chicago: Univ. of Chicago Press. 816 pp.

Guyer, Craig, and Maureen A. Donnelly. 2005. *Amphibians and Reptiles of La Selva, Costa Rica, and the Caribbean Slope*. Berkeley: Univ. of California Press. 299 pp.

Hall, E. Raymond. 1981. *The Mammals of North America*. Vols. 1 and 2. New York: John Wiley & Sons. 1181 pp.

Heaney, Larry R. 1983. Red-tailed Squirrel (*Sciurus granatensis*). In *Costa Rican Natural History*, ed. Daniel H. Janzen, 489–490. Chicago: Univ. of Chicago Press. 816 pp.

Henderson, Carrol L. 1970. Fish and Wildlife Resources in Costa Rica, with Notes on Human Influences. Master of Forest Resources thesis, Univ. of Georgia, Athens. 340 pp.

Holdridge, Leslie R. 1967. *Life Zone Ecology*. San José, Costa Rica: Tropical Science Center. 206 pp.

Holland, Jennifer S. 2009. Vanishing Amphibians. *National Geographic* 215(4): 138–153.

Horwich, Robert H., and Jonathan Lyon. 1990. *A Belizean Rain Forest: The Community Baboon Sanctuary*. Gays Mills, Wisc.: Orang-utan Press. 420 pp.

INICEM. 1998. Costa Rica: Datos e Indicadores Básicos. Costa Rica at a Glance. Miami: INICEM Group. Booklet. 42 pp.

Janzen, Daniel H. 1983. White-tailed Deer (*Odocoileus virginianus*). In *Costa Rican Natural History*, ed. Daniel H. Janzen, 481–483. Chicago: Univ. of Chicago Press. 816 pp.

———. 1990. Costa Rica's New National System of Conserved Wildlands. Mimeographed report. 15 pp.

———. 1991. How to Save Tropical Biodiversity: The National Biodiversity Institute of Costa Rica. *American Entomologist* 36(3): 159–171.

———, ed. 1983. *Costa Rican Natural History*. Chicago: Univ. of Chicago Press. 816 pp.

Joyce, Christopher. 1994. *Earthly Goods: Medicine-hunting in the Rainforest*. Boston: Little Brown. 304 pp.

Kaufmann, John H. 1983. White-nosed Coati (*Nasua narica*). In *Costa Rican Natural History*, ed. Daniel H. Janzen, 478–480. Chicago: Univ. of Chicago Press. 816 pp.

Kohl, Jon. 1993. No Reserve Is an Island. *Wildlife Conservation* 96(5): 74–75.

Kubicki, Brian. 2000. The Centrolenidae Family of Neotropical Glass Frogs. *Reptiles* 8(8): 48–60.

———. 2004. *Ranas de hoja de Costa Rica* (Leaf Frogs of Costa Rica). Santo Domingo de Heredia, Costa Rica: Editorial INBIO. 117 pp.

———. 2004. Rediscovery of *Hyalinobatrachium chirripoi* (Anura: Centrolenidae) in Costa Rica. *Rev. Biol. Trop.* 52(1): 215–218.

———. 2006. Rediscovery of the green-striped glass frog *Hyalinobatrachium talamancae* (Anura: Centrolenidae) in Costa Rica. *Brenesia* 66: 25–30.

———. 2007. *Ranas de vidrio de Costa Rica* (Glass Frogs of Costa Rica). Santo Domingo de Heredia, Costa Rica: Editorial INBIO. 304 pp.

———. 2008. Amphibian diversity in Guayacán, Limón Province, Costa Rica. *Brenesia* 69: 35–42.

LaVal, Richard K., and Bernal Rodríguez-Herrera. 2002. *Murciélagos de Costa Rica* (*Costa Rica Bats*). Santo Domingo de Heredia, Costa Rica: INBIO. 320 pp.

Leenders, Twan. 2001. *A Guide to Amphibians and Reptiles of Costa Rica*. Miami: Zona Tropical Publications. 305 pp.

Leopold, Aldo S. 1959. *Wildlife of Mexico: The Game Birds and Mammals*. Berkeley: Univ. of California Press. 568 pp.

Lewin, Roger. 1988. Costa Rican Biodiversity. *Science* 242: 1637.

Lewis, Thomas A. 1989. Daniel Janzen's Dry Idea. *International Wildlife* 19(1): 30–36.

Lubin, Yael D. 1983. Tamandua (*Tamandua mexicana*). In *Costa Rican Natural History*, ed. Daniel H. Janzen, 494–496. Chicago: Univ. of Chicago Press. 816 pp.

Macdonald, David, ed. 1984. *All the World's Animals: Primates*. New York: Torstar Books. 160 pp.

Market Data. 1993. Costa Rica: Datos e Indicadores Básicos. Costa Rica at a Glance. San José, Costa Rica. 36 pp.

McDiarmid, Roy W. 1983. *Centrolenella fleischmanni*. In *Costa Rican Natural History*, ed. Daniel H. Janzen, 389–390. Chicago: Univ. of Chicago Press. 816 pp.

McPhaul, John. 1988. Peace, Nature: C. R. Aims. *Tico Times* 32(950): 1, 21.

Meña Moya, R. A. 1978. *Fauna y caza en Costa Rica*. San José, Costa Rica: Litografía e Imprenta LIL. 255 pp.

Meza Ocampo, Tobías A. 1988. *Areas Silvestres de Costa Rica*. San Pedro, Costa Rica: Alma Mater. 112 pp.

Murillo, Katiana. 1999. Ten Years Committed to Biodiversity. *Friends in Costa Rica* 3: 23–25.

Norman, David. 1998. Common Amphibians of Costa Rica. Heredia, Costa Rica: Published by the author. 96 pp.

Pariser, Harry S. 1996. *Adventure Guide to Costa Rica*. 3d ed. Edison, N.J.: Hunter. 546 pp.

Pistorius, Robin, and Jeroen van Wijk. 1993. Biodiversity Prospecting: Commercializing Genetic Resources for Export. *Biotechnology and Development Monitor* 15: 12–15.

Pounds, J. A., Michael P. L. Fogden, Jay M. Savage, and G. C. Gorman. 1997. Tests of Null Models for Amphibian Declines on a Tropical Mountain. *Conservation Biology* 11(6): 1307–1322.

Pratt, Christine. 1999. Tourism Pioneer Wins Award, Hosts Concorde. *Tico Times* 43(1507): 4.

Reid, Fiona A. 1997. *A Field Guide to the Mammals of Central America and Southeast Mexico.* New York: Oxford Univ. Press. 334 pp.

Rich, Pat V., and T. H. Rich. 1983. The Central American Dispersal Route: Biotic History and Paleogeography. In *Costa Rican Natural History,* ed. Daniel H. Janzen, 12–34. Chicago: Univ. of Chicago Press. 816 pp.

Robinson, Douglas C. 1983. Malachite Lizard (*Sceloporus malachiticus*). In *Costa Rican Natural History,* ed. Daniel H. Janzen, 421–422. Chicago: Univ. of Chicago Press. 816 pp.

Rowe, Noel. 1996. *The Pictorial Guide to the Living Primates.* East Hampton, N.Y.: Pogonias Press. 263 pp.

Rudloe, Anne, and Jack Rudloe. 1994. Sea Turtles: In a Race for Survival. *National Geographic* 185(2): 94–121.

Sandlund, Odd Terje. 1991. Costa Rica's INBIO: Towards Sustainable Use of Natural Biodiversity. Norsk Institutt for Naturforskning. Notat 007. Trondheim, Norway. Report. 25 pp.

Savage, Jay M. 2002. *The Amphibians and Reptiles of Costa Rica.* Chicago: Univ. of Chicago Press. 934 pp.

Scott, Norman J. 1983. Red-eyed Tree Frog (*Agalychnis callidryas*). In *Costa Rican Natural History,* ed. Daniel H. Janzen, 374–375. Chicago: Univ. of Chicago Press. 816 pp.

———. Litter Toad (*Bufo haematiticus*). In *Costa Rican Natural History,* ed. Daniel H. Janzen, 385. Chicago: Univ. of Chicago Press. 816 pp.

———. Bransford's Litter Frog (*Eleutherodactylus bransfordii*). In *Costa Rican Natural History,* ed. Daniel H. Janzen, 399. Chicago: Univ. of Chicago Press. 816 pp.

———. Smoky Jungle Frog (*Leptodactylus pentadactylus*). In *Costa Rican Natural History,* ed. Daniel H. Janzen, 405–406. Chicago: Univ. of Chicago Press. 816 pp.

———. Vine Snake (*Oxybelis aeneus*). In *Costa Rican Natural History,* ed. Daniel H. Janzen, 410–411. Chicago: Univ. of Chicago Press. 816 pp.

Seifert, Robert P. 1983. Eyelash Viper (*Bothrops schlegelii*). In *Costa Rican Natural History,* ed. Daniel H. Janzen, 384–385. Chicago: Univ. of Chicago Press. 816 pp.

Sekerak, Aaron D. 1996. *A Travel and Site Guide to Birds of Costa Rica.* Edmonton, Alberta: Lone Pine. 256 pp.

Skutch, Alexander F. 1971. *A Naturalist in Costa Rica.* Gainesville: University Press of Florida. 378 pp.

———. 1984. Your Birds in Costa Rica. Santa Monica, Calif.: Ibis. Brochure. 8 pp.

Sowls, L. K. 1983. Collared Peccary (*Tayassu tajacu*). In *Costa Rican Natural History,* ed. Daniel H. Janzen, 497–498. Chicago: Univ. of Chicago Press. 816 pp.

Sun, Marjorie. 1988. Costa Rica's Campaign for Conservation. *Science* 239: 1366–1369.

Sunquist, Fiona. 1986. The Secret Energy of the Sloth. *International Wildlife* 16(1): 4–11.

Tangley, Laura. 1990. Cataloging Costa Rica's Diversity. *BioScience* 40(9): 633–636.

Timm, Robert M., D. E. Wilson, B. L. Clauson, R. K. LaVal, and C. S. Vaughan. 1989. *Mammals of the La Selva–Braulio Carrillo Complex, Costa Rica.* North American Fauna Series, 75. Washington, D.C.: U.S. Dept. of the Interior, Fish and Wildlife Service. 162 pp.

Tyson, Peter. 1997. High-tech Help for Ancient Turtles. *MIT's Technology Review* 10(8): 54–60.

Ugalde, Alvaro F., and María Luisa Alfaro. 1992. Financiamiento de la Conservación en los Parques Nacionales y Reservas Biológicas de Costa Rica. Speech presented at the IV Congreso Mundial de Parques Nacionales, Caracas, Venezuela, February. Mimeographed copy. 16 pp.

Van Devender, R. Wayne. 1983. Basiliscus Lizard (*Basiliscus basiliscus*). In *Costa Rican Natural History,* ed. Daniel H. Janzen, 379–380. Chicago: Univ. of Chicago Press. 816 pp.

Wainwright, Mark. 2007. *The Mammals of Costa Rica: A Natural History Field Guide.* Ithaca, N.Y.: Zona Tropical Publication, Cornell Univ. Press. 454 pp.

Walls, Jerry G. 1994. *Jewels of the Rainforest: Poison Frogs of the Family Dendrobatidae.*

Neptune City, N.J.: TFH Publications. 288 pp.

———. 1996. *Red-eyes and Other Leaf Frogs.* Neptune City, N.J.: TFH Publications. 64 pp.

Witt, Matthew J., B. Baert, A. C. Broderick, A. Formia, J. Fretey, A. Gibudi, C. Moussounda, G. A. M. Mounguengui, S. Ngouessono, R. J. Parnell, D. Roumet, G. Sounguet, B. Verhage, A. Zogo, and B. J. Godley. 2009. Aerial Surveying of the World's Largest Leatherback Turtle Rookery: A More Effective Methodology for Large-scale Monitoring. *Biological Conservation* 142(8): 1719–1727.

Zug, George. 1983. Giant Toad (*Bufo marinus*). In *Costa Rican Natural History,* ed. Daniel H. Janzen, 386–387. Chicago: Univ. of Chicago Press. 816 pp.

Zúñiga Vega, Alejandra. 1991. Archivo de riqueza natural. *La Nación,* Section B, Viva, February 4.

———. 1991. Estudios de los manglares. *La Nación,* Section B, Viva, February 4.

APPENDICES

APPENDIX A: COSTA RICAN CONSERVATION ORGANIZATIONS, RESEARCH STATIONS, BIRDING GROUPS, AND BIRD INFORMATION SOURCES

Asociación Ornitológica de Costa Rica: Apdo 2289-1002, San José, Costa Rica. E-mail: chidalgo@una.ac.cr or drivera@una.ac.cr. Newsletter: *Zeledonia*. Office address: Avenida 6, Calles 21 y 25, Casa 2194. Telephone: 506-2256-9587.

Association for the Conservation of Nature (ASCONA): Apdo 83790-1000, San José, Costa Rica. Telephone: 506-2233-3188.

Birding Club of Costa Rica: Web newsletter *Tico Tweeter* reports on recent and upcoming birding trips and birding discoveries at various sites and lodges. www.ticotours.home.att.net/Tweeter.

Caribbean Conservation Corporation: Gainesville, Florida. Telephone: 1-800-678-7853. www.cccturtle.org.

Corcovado Foundation: A nonprofit organization dedicated to protection of the rainforest and wildlife in Corcovado National Park, where there have been recent problems with poaching of endangered wildlife species. www.corcovadofoundation.org.

Costa Rica Birding Trail: A new effort in the northern region of Costa Rica to promote birding along a trail of lodges and protected areas that provide a stimulating variety of habitats. www.costaricanbirdroute.com.

Gone Birding: Web newsletter about birding and birding activities in Costa Rica, authored by expert birder Richard Garrigues. Google "gone birding newsletter."

Great Green Macaw Research and Conservation Project: Cooperative project headed by the Tropical Science Center and dedicated to the protection and restoration of the Great Green Macaw in the Caribbean lowlands of Costa Rica. www.lapaverde.or.cr/lapa; e-mail: lapa@cct.or.cr.

Henderson Birding: www.hendersonbirding.com. The author's Web site, where updates, new information, and corrections to the *Field Guide to the Wildlife of Costa Rica* will be posted. Includes Costa Rica birding information, trip tips, preparation checklist, and itineraries for future birding trips.

Las Cruces Biological Station and Wilson Botanical Garden, OTS: Apdo 73, 8257 San Vito, Coto Brus, Costa Rica. Telephone: 506-2773-4004; e-mail: lcruces@hortus.ots.ac.cr.

La Selva Biological Field Station, OTS: Apdo 53-3069, Puerto Viejo de Sarapiquí, Heredia, Costa Rica. Telephone: 506-2766-6565; e-mail: laselva@sloth.ots.ac.cr.

Leatherback Trust: Leatherback Turtle conservation organization. www.leatherback.org.

Los Cusingos Neotropical Bird Sanctuary: The former home of Dr. Alexander Skutch, now managed by the Tropical Science Center. Apdo 8-3870-1000, San José, Costa Rica. Telephone: 506-2253-3276; e-mail: cecitrop@sol.racsa.co.cr, reservations cusingos-reservation@cct.or.cr. www.cct.or.cr/cusingos.

Monteverde Conservation League: Apdo 10165-1000, San José, Costa Rica. Manages the Children's Eternal Rainforest. Telephone: 506-2645-5003; e-mail: acmmcl@sol.racsa.co.cr. www.acmcr.org; also www.monteverdeinfo.com.

National Biodiversity Institute (INBIO): Apdo 22-3100, Santo Domingo, Heredia, Costa Rica. National organization dedicated to the creation of a comprehensive inventory of all of Costa Rica's plant and wildlife species; with biological collections now exceeding 3.5 million specimens and many excellent publications. Telephone: 506-2507-8100. www.inbio.ac.cr.

National Parks Foundation (Fundación de Parques Nacionales): Apdo 236-1002, San José, Costa Rica. Telephone: 506-2222-4921 or 506-2223-8437.

Organization for Tropical Studies, Inc.: North American Headquarters, Box 90630, Durham, NC 27708-0630. www.ots.duke.edu; also www.ots.ac.cr. Costa Rican office: Apdo 676, 2050 San Pedro do Montes de Oca, San José,

Costa Rica. Telephone: 506-2240-6696; e-mail: oet@cro.ots.ac.cr.

Rainforest Action Network: 221 Pine Street, Suite 500, San Francisco, CA 94104. Telephone: 415-398-4404. www.ran.org.

Rincón Rainforest: A protected tropical forest of 13,838 acres in northern Costa Rica. Donations to the Guanacaste Dry Forest Conservation Fund help to save Costa Rica's biodiversity; tax-deductible donations can be matched from conservation foundations and will be used 100 percent for land acquisition (no overhead or administrative charges). Contact Dr. Daniel H. Janzen for further details; djanzen@sas.upenn.edu. Donations made out to the Guanacaste Dry Forest Conservation Fund can be sent to Prof. Daniel H. Janzen, Dept. of Biology, 415 South University Ave., University of Pennsylvania, Philadelphia, PA 19104. http://janzen.sas.upenn.edu/RR/rincon_rainforest.htm.

Tirimbina Rainforest Center: A tropical science research and tourism center initiated by the Milwaukee Public Museum. Telephone: 506-2761-1579 or 506-2761-0055; e-mail: info@tirimbina.org. www.envirolink.org.

APPENDIX B: WILDLIFE TOURISM SITES AND FIELD STATIONS OF COSTA RICA

The wildlife tourism sites described here and portrayed in the map include seventy-six sites that have been visited by the author in Costa Rica, many of which are referred to in this book. In the following key, the site code from Figure 9 is followed by the name of the site, its biological zone and elevation, its coordinates, a brief description, and contact information. Abbreviations: BR, Biological Reserve; NP, National Park; NWR, National Wildlife Refuge; OTS, Organization for Tropical Studies; PAH, Pan-American Highway.

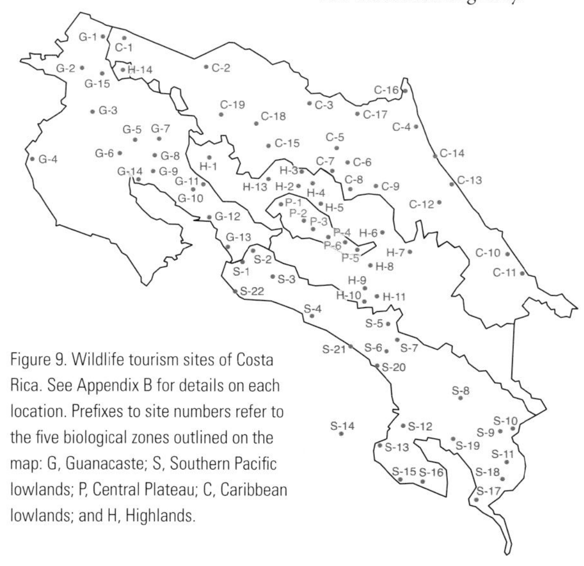

Figure 9. Wildlife tourism sites of Costa Rica. See Appendix B for details on each location. Prefixes to site numbers refer to the five biological zones outlined on the map: G, Guanacaste; S, Southern Pacific lowlands; P, Central Plateau; C, Caribbean lowlands; and H, Highlands.

GUANACASTE REGION

G-1: Los Inocentes Ranch: Tropical moist forest. Elev. 750'. Lat. 11°02.50'N, long. 85°30.00'W. Address: P.O. Box 228-3000, Heredia, Costa Rica. This lodge is now closed to tourism.

G-2: Santa Rosa NP: Tropical dry forest/premontane moist forest. Elev. 1350'. Lat. 10°51.50'N, long. 85°36.50'W. This park contains 181,186 acres and is an excellent example of tropical dry forest and premontane moist forest, as well as gallery forests. The Olive Ridley Sea Turtle nesting beaches of Nancite are within this park. Telephone: Santa Rosa NP, 506-2666-5051; Guanacaste Conservation Area, 506-2666-4740.

G-3: Liberia, road to Tamarindo: Tropical dry forest. Elev. 100'. From lat. 10°37.50'N, long. 85°27.00'W, to lat. 10°18.60'N, long. 85°55.00'W. Many species of the tropical dry forest can be spotted along this road en route to see the Leatherback Turtles at Tamarindo.

G-4: Tamarindo, Playa Grande, Las Baulas NP, and Sugar Beach: Tropical dry forest. Elev. Sea level. Lat. 10°18.60'N, long. 85°55.00'W. Las Baulas NP, which covers 1,364 acres, protects the nesting beaches of the Leatherback Turtle at Playa Grande. Mangrove lagoons and beaches in the vicinity are important wintering sites for shorebirds, wading birds, and migratory wildlife. Tropical dry forests of the area provide opportunities for viewing howler monkeys and other upland wildlife. Tamarindo NWR and Las Baulas NP, telephone: 506-2653-0470. Turtle tours available October through March at El Mundo de las Tortugas, 506-2653-0471, at Playa Grande. Telephone: Hotel Bula Bula, 506-2653-0975, www.hotelbulabula.com; Hotel Las Tortugas, 506-2653-0423, www.cool.co.cr/usr/turtles; Hotel Villa Baula, 506-2653-0644; www.hotelvillabaula.com.

G-5: Lomas Barbudal BR: Tropical dry forest. Elev. 100'. Lat. 10°26.20'N, long. 85°16.00'W. A reserve of 5,631 acres that provides an excellent example of riparian forest within the Guanacaste region. Elegant Trogons, Scrub Euphonias, and Long-tailed Manakins are among the featured wildlife there.

G-6: Palo Verde NP. Tropical dry forest. Elev. 30'. Lat. 10°22.21'N, long. 85°11.84'W. One of the best examples of both dry forest and tropical wetlands. Includes the area formerly designated as the Dr. Rafael Lucas Rodríguez Caballero NWR, with 45,511 acres of tropical dry forest, riparian forest, and marshes. It is an important wintering site for migratory waterfowl and local Black-bellied Whistling-Ducks and Muscovy Ducks; also home to Jabiru Storks, Scarlet Macaws, and Snail Kites. This is an important research and education site for the OTS and for the National Biodiversity Institute. Palo Verde OTS Biological Station, telephone: 506-2524-0607; www.threepaths.co.cr. Reservations: edu.travel@ots.ac.cr.

G-7: La Pacífica (Hotel Hacienda La Pacífica) and Cañas: Tropical dry forest. Elev. 150'. Lat. 10°27.21'N, long. 85°07.68'W. This lodge, on a 6,548-acre ranch and private forest reserve formerly known as Finca La Pacífica, is an excellent place to stay when visiting locations like Guanacaste, Santa Rosa, and Palo Verde NPs and Lomas Barbudal BR. Address: Apdo 8, 5700 Cañas, Guanacaste, Costa Rica; www.pacificacr.com. Telephone: 506-2669-6050; e-mail: pacifica@racsa.co.cr. On the east boundary of this property is Las Pumas, a wild cat rescue and rehabilitation center. All six of Costa Rica's wild cats can be observed there. Telephone: 506-2669-6044. Donations are appreciated.

G-8: Estancia Jiménez Núñez and lagoons: Tropical dry forest. Elev. 250'. Lat. 10°20.55'N, long. 85°08.69'W. Private ranch with large man-made lagoons with many waterbirds; good raptor viewing on the road from the PAH west to this ranch. Get permission from the guard at the entrance to see the lagoons.

G-9: Hacienda Solimar: Tropical dry forest. Elev. 100'. Lat. 10°15.58'N, long. 85°09.40'W. Excellent dry forest and riparian forest, with exceptional wetland wildlife, including Roseate Spoonbills, Snail Kites, nesting Jabiru Storks, Boat-billed Herons, Bare-throated Tiger-Herons, and crocodiles. Owners have made significant improvements to this 5,000-acre ranch to accommodate wildlife tourism. Telephone, at ranch: 506-2669-0281; e-mail: solimar@racsa.co.cr.

G-10: Río Lagarto bridge and farm lagoon: Premontane moist forest. Elev. 140'. Lat. 10°09.76'N, long. 84°54.93'W. This is a farm pond just off the PAH north of the bridge over the Río Lagarto. It is at Ganadería Avancari. Black-bellied Whistling-Ducks, Least Grebes, Purple Gallinules, and Northern Jacanas are regularly observed along the road by the marsh.

G-11: Pulpería La Pita and lowlands to Monteverde: Tropical moist forest. Elev. 700'. From lat. 10°10.09'N, long. 84°54.38'W to lat. 10°18.00'N, long. 84°49.20'W. This is the road from the turnoff from the PAH by Río Lagarto, past a small store known as Pulpería La Pita, and through mixed pasture and woodland en route to Monteverde. Wildlife includes species of the Guanacaste dry forest, like White-throated Magpie-Jays, Crested Bobwhites, Rufous-naped Wrens, and Long-tailed Manakins.

G-12: Puntarenas, Hotel Tioga, and Playa Doña Ana: Premontane wet forest. Elev. Sea level. Lat. 9°58.46'N, long. 84°50.34'W. The lagoons and beaches of Puntarenas and nearby Playa Doña Ana provide excellent areas to observe shorebirds, wading birds, White-winged Doves, frigatebirds, cormorants, Black Skimmers, Anhingas, terns, and gulls. Hotel Tioga, with its downtown beachfront location, provides easy access to the beach, local wetlands, and nearby wildlife sites like Carara NP. Address: P.O. Box 96-5400, Puntarenas, Costa Rica; www.hoteltioga.com. Telephone: in San José, 506-2255-3115; in Puntarenas, 506-2661-0271.

G-13: Bajamar: Tropical dry forest. Elev. Sea level. Lat. 9°50.50'N, long. 84°40.50'W. This area of dry forest and mangrove lagoons near the coast represents the southern range limit for dry forest birds of the Guanacaste region.

G-14: La Ensenada Lodge: Tropical dry forest. Elev. Sea level to 250'. Lat. 10°08.304'N, long. 85°02.394'W. A privately owned national wildlife reserve, this 939-acre ranch is a wildlife mecca at the head of the Gulf of Nicoya. It contains tropical dry forest, pastures, mangrove lagoons, shoreline habitat, wetlands, and commercial salt ponds that attract great varieties of shorebirds and dry forest wildlife. This ranch is used by Three-wattled Bellbirds from December through February, as well as Turquoise-browed Motmots and White-throated Magpie-Jays. Boat tours in nearby mangrove lagoons provide the chance to see the Mangrove Vireo, Mangrove Hummingbird, Mangrove Cuckoo, and Mangrove race of the Yellow Warbler. This is one of the only places in Costa Rica where the Northern Potoo can be encountered at night. www.laensenada.net. Telephone: 506-2289-6655 or 506-2289-7443; e-mail: la_ensenada@yahoo.com.

G-15: Hotel Borinquen Mountain Resort and Thermal Spa: Tropical moist forest/premontane moist forest transition. Elev. 2,500'. Lat. 10°48.704'N, long. 84°04.965'W. This resort and spa features thermal springs and also provides good birding opportunities on the grounds and vicinity, including access to nearby Rincón de la Vieja NP. Wildlife is typical of the tropical dry forest but also features some species of the Caribbean slope. There are Turquoise-browed Motmots, Crested Caracaras, Red-lored Parrots, White-fronted Parrots, Keel-billed Toucans, White-throated Magpie-Jays, Double-striped Thick-knees, and Plain-capped Starthroats. www.borinquenresort.com. Telephone: 506-2690-1900; e-mail: borinque@racsa.co.cr.

SOUTHERN PACIFIC LOWLANDS

S-1: Carara NP, Río Tárcoles estuary, Tárcol Lodge, Villa Caletas, Crocodile Jungle Safari, and Villa Lapas: Carara NP: Premontane moist forest. Elev. Sea level–100'. Lat. 9°47.72'N, long. 84°36.16'W. Carara NP is an excellent reserve, covering 12,953 acres, that has wildlife of both the Guanacaste tropical dry forest and tropical wet forests of the southern Pacific lowlands. It is one of the best places to observe Scarlet Macaws and crocodiles in the country. Telephone: Carara NP, 506-2383-9953. Tárcol Lodge, at the mouth of the Río Tárcoles, is now closed. Villa Caletas: Tropical moist forest. Elev. Sea level–150'. Lat. 9°41.207'N, long. 84°39.597'W. Villa Caletas is an excellent seaside resort that provides easy access to Carara NP and boat tours in the Río Tárcoles mangrove lagoons. Many notable wildlife species can be seen right on the grounds, including Zone-tailed Hawks, parrots, chachalacas, and hummingbirds. Address: P.O. Box 12358-1000, San José, Costa Rica; www.villacaletas.com. Telephone: Reservations, 506-2637-0606; Hotel, 506-2637-0505; e-mail: reservations@villacaletas.com. Crocodile Jungle Safari Tour: Tropical dry forest. Elev. Sea level. Lat. 9°46.930'N, long. 84°38.187'W. Crocodile Jungle Safari offers outstanding boat tours in the mangrove lagoons and estuaries of the Río Tárcoles near Carara NP. The guides are exceptional at spotting and identifying wildlife, and they are successful in locating a variety of birds and providing great photo opportunities. This tour company provides good opportunities for viewing the crocodiles. They are recommended because they do not feed or habituate the crocodiles to their presence like some tour operators. Address: P.O. Box 1542, San Pedro, Costa Rica; www.costaricanaturetour.com. Telephone: 506-2637-0338; e-mail: crocodile@costaricanaturetour.com. Hotel Villa Lapas: Tropical moist forest. Elev. approx. 150'. Lat. 9°45.368'N, long. 84°36.573'W. The grounds of this resort offer some outstanding birding and easy access to nearby Carara NP and the Río Tárcoles mangrove lagoons. Trogons, owls, tiger-herons, and even Scarlet Macaws can be observed on the property. The skywalk facility in the adjacent forest provides outstanding opportunities to see wildlife of the moist and dry forest, including Long-tailed Manakins, hummingbirds, parrots, tanagers, and trogons. Address: P.O. Box 419-4005, Heredia, Costa Rica; www.villalapas.com. Telephone: 506-2637-0232; e-mail: info@villalapas.com. Hotel Xandari by the Beach: South of Carara; www.hotelxandari.com. Telephone: 506-2778-7070.

S-2: Orotina, road to lowlands approaching Carara: Tropical moist forest. Elev. 500'. Lat. 9°51.80'N, long. 84°34.00'W. This route includes the famous city park in Orotina where a pair of Black-and-White Owls and an introduced Two-toed Sloth have lived for many years.

S-3: Parrita, road to Puriscal: Tropical moist forest. Site 1: Elev. 1,600'. Lat. 9°42.46'N, long. 84°24.22'W. Site 2: Elev. 2,000'. Lat. 9°43.90'N, long. 84.23.64'W.

S-4: Manuel Antonio NP, Rancho Casa Grande, Damas Island mangrove tours, and Quepos: Tropical wet forest. Elev. Sea level. Lat. 9°22.94'N, long. 84°08.62'W. Manuel Antonio NP covers 4,015 acres of land and 135,905 acres of ocean. It provides excellent opportunities to see squirrel monkeys, agoutis, white-faced monkeys, ctenosaurs, butterflies, and many species of the southern Pacific lowlands. Telephone: 506-2777-3130, 506-2777-1646; toll-free reservations: 1-888-790-5264; e-mail: osrap@minae

.go.cr. Foresta Resort Rancho Casa Grande: Premontane wet forest. Elev. approx. 100–250'. Lat. 9°26.414'N, long. 84°08.185'W. Hotel Rancho Casa Grande is an excellent lodge to stay at while exploring the Quepos area, including Manuel Antonio NP and the nearby mangrove lagoons. There are some outstanding trails for birding on the 180 acres. It is a good place to see endangered squirrel monkeys, tityras, Streaked Flycatchers, and other wildlife of the southern Pacific lowlands. Address: P.O. Box 618-2010 Zapote, San José, Costa Rica; ranchocasagrande.com. Telephone: 506-2777-3130, 506-2777-1646; e-mail: hotelrcg@sol.racsa.co.cr. Damas Island mangrove lagoons: Iguana Tours (Jorge's Mangrove Tours) in Quepos can arrange boating tours to see the wildlife of the mangrove lagoons near Damas Island. It is one of the best places to look for the rare Silky Anteater, Mangrove Hummingbird, Mangrove Vireo, and the Mangrove subspecies of the Yellow Warbler. www.iguanatours.com. Telephone: 506-2777-1262; e-mail: iguana@racsa.co.cr.

S-5: Talari Mountain Lodge near San Isidro del General: Premontane wet forest. Elev. 2,800'. Lat. 9°24.14'N, long. 83°40.12'W. This rustic lodge near San Isidro has great wildlife viewing opportunities on the grounds. It is easy to observe seventy species in a morning of birding there. This is an excellent place to see the Slaty Spinetail, Pearl Kite, and Red-legged Honeycreeper. Many birds come to the feeders. Address: Talari Albergue de Montaña, Rivas, San Isidro del General, Apdo 517-8000, Pérez Zeledón, Costa Rica; www.talari.co.cr. Telephone and fax: 506-2771-0341; e-mail: talaricostarica@gmail.com. This is a good place to stay if visiting Los Cusingos, the former home of famous ornithologist Alexander Skutch, now managed by the Tropical Science Center in San José (see details for site S-7).

S-6: La Junta de Pacuares resort on Río General: Tropical moist forest. Elev. 2,200'. Lat. 9°16.51'N, long. 83°38.33'W. A variety of wildlife can be seen along the river at this site, including Gray-headed Chachalacas.

S-7: Los Cusingos, San Isidro del General, City Lagoons, and Hotel del Sur: Tropical moist forest. San Isidro del General: Elev. 2,200'. Lat. 9°20.40'N, long. 83°28.00'W.; City sewage lagoons: Elev. 2,000'. Lat. 9°22.25'N, long. 83°41.80'W.; Los Cusingos: Elev. 2,500'. Lat. 9°19.10'N, long. 83°36.75'W. Los Cusingos Neotropical Bird Sanctuary is the former homestead of the late Dr. Alexander and Pamela Skutch. It is now managed by the Tropical Science Center. Los Cusingos is an excellent remnant forest reserve where it is possible to make a day trip to see Bay-headed Tanagers, Gray-headed Chachalacas, Speckled Tanagers, White-breasted Wood-Wrens, and other wildlife of the southern Pacific moist forest. Make arrangements for visits with the Tropical Science Center, Apdo 8-3870-1000, San José, Costa Rica; www.cct.or.cr (Refugio Los Cusingos). Telephone: 506-2253-3267; e-mail: cct@cct.or.cr. Hotel del Sur in San Isidro del General and Talari Mountain Lodge (site S-5) provide convenient places to stay. Address: Hotel del Sur, P.O. Box 4-8000, San Isidro del General, Costa Rica; Google "Hotel del Sur, Costa Rica." Telephone: 506-2771-3033; e-mail: reservas@hoteldelsur.com.

S-8: Río Térraba bridge crossing: Tropical moist forest. Elev. 200'. Bridge over Río Térraba: Lat. 9°00.20'N, long. 83°13.20'W. This bridge crossing is well known for the huge crocodiles that can be seen there, as well as herons and egrets.

S-9: Wilson Botanical Garden at San Vito–OTS Las Cruces Biological Field Station: Premontane rainforest. Elev. 3,900'. Wilson Botanical Garden, lat. 8°49.61'N, long. 82°57.80'W. This OTS field station has cabins for tourists and an excellent trail system. There are regionally endemic birds on the property, and many birds come to the feeders there. Address: Apdo 73-8257, San Vito, Costa Rica; main OTS address: Apdo 676-2050 San Pedro de Montes Oca, Costa Rica; www.threepaths.com. Telephone: in USA, 1-919-684-5774; in San José, 506-2524-0607; reservations in San José, 506-2524-0628; e-mail: edu.travel@ots.ac.cr. Wetland species can be observed nearby at the lagoons by the airport (for a fee; this is on private property) and at Los Contaros, a private nature park and wetland with a gift shop owned by Gail Hewson. It has indigenous crafts of the Guaymi community and is on the outskirts of San Vito, near the Wilson Botanical Garden.

S-10: Sabalito, road from San Vito: Premontane wet forest. Elev. 2,200'. From lat. 8°49.61'N, long. 82°57.80'W, to lat. 8°49.80'N, long. 82°53.80'W. When exploring the San Vito area, along this road is a good place to look for southern bird specialties like the Bran-colored Flycatcher, Pearl Kite, Masked (Chiriquí) Yellowthroat, and Crested Oropendola.

S-11: Paso Canoas: Premontane wet forest. Elev. 300'. From lat. 8°32.00'N, long. 82°50.30'W, to lat. 8°49.61'N, long. 82°57.80'W. Paso Canoas is the town in southern Costa Rica where the PAH enters Panama. Wildlife of the area is typical of the southern Pacific lowlands.

S-12: Sierpe, on the Río Térraba, to Drake Bay: Premontane wet forest.

Elev. Sea level. From Sierpe at lat. 8°51.50'N, long. 83°28.20'W, to the Río Sierpe estuary at lat. 8°46.50'N, long. 83°38.00'W. Local lodging includes Río Sierpe Lodge, where boat trips are available to Corcovado NP and Caño Island BR. Telephone: 506-2283-5573; e-mail: vsftrip@racsa.co.cr.

S-13: Drake Bay Wilderness Resort, northwest end of Corcovado NP: Tropical wet forest. Elev. Sea level. Lat. 8°41.80'N, long. 83°41.00'W. www.drakebay.com. Telephone: 506-2770-8012 (also fax), 506-2384-4107; in San José, 506-2256-7394; at resort, 506-2371-3437; e-mail: hdrake@sol.racsa.co.cr.

S-14: Caño Island BR: Tropical wet forest. Elev. Sea level. Lat. 8°43.00'N, long. 83°53.00'W. This island, six miles from the Osa Peninsula, includes 741 acres. It is possible to see Humpback Whales while en route between the mainland and the island. Arrange for visits with local lodges like Río Sierpe Lodge, Drake Bay Wilderness Resort, Aguila de Osa Inn, La Paloma Lodge, or Marenco Lodge.

S-15: Sirena Biological Station, Corcovado NP: Tropical wet forest. Elev. Sea level. Lat. 8°78.74'N, long. 83°35.81'W. This biological station is one of the best examples of remote, wild rainforest in Costa Rica. There are significant populations of Scarlet Macaws, tapirs, White-lipped Peccaries, Great Curassows, jaguars, pumas, and other species characteristic of tropical wet forests. Accessible by air or on foot by hiking along the beach from La Leona or San Pedrillo.

S-16: Corcovado Lodge Tent Camp, Carate, and southeast end of Corcovado NP: Tropical wet forest. Elev. Sea level. Lat. 8°26.88'N, long. 83°28.97'W. This excellent tent camp lodge provides great access for wildlife viewing on the beachfront property, along trails in the rainforest behind the lodge, and in nearby Corcovado NP. A wildlife tower allows viewing of wildlife in the forest canopy. Wildlife includes Scarlet Macaws, parrots, King Vultures, and many raptors. Address: Costa Rica Expeditions, P.O. Box 6941-1000, San José, Costa Rica; www.costaricaexpeditions.com. Telephone: 506-2257-0766, 506-2222-0333; e-mail: costaric@expeditions.co.cr. Similar wildlife can be observed at Marenco Beach and Rainforest Lodge (506-2770-8002; www.marencolodge.com), Bosque del Cabo (506-2735-5206; www.bosquedelcabo.com), Luna Lodge (506-2380-5036; www.lunalodge.com), La Paloma Lodge (506-2239-2801; www.lapalomalodge.com), and Lapa Ríos (506-2735-513; www.laparios.com).

S-17: Tiskita Jungle Lodge: Tropical wet forest. Elev. Sea level to 200'. Lat. 8°21.48'N, long. 83°8.05'W. A 400-acre private forest reserve and tropical fruit experimental field station, this is one of the best places in Costa Rica to see squirrel monkeys and White Hawks. Many tanagers and honeycreepers are present because of the variety of fruiting trees on the grounds. Address: Costa Rica Sun Tours, P.O. Box 13411-1000, San José, Costa Rica; www.tiskita.com. Telephone: 506-2296-8125.

S-18: Road from San Vito to Paso Canoas: Premontane moist forest. Elev. 900'. Lat. 8°44.50'N, long. 82°56.90'W. The highway from San Vito de Java to Paso Canoas provides a good opportunity for viewing wildlife of the southern Pacific lowlands, including Blue-headed Parrots, squirrel monkeys, and relatively new arrivals in Costa Rica like the Pearl Kite and Crested Oropendola. There is a colony of Crested Oropendolas nesting along the highway near Villa Neily.

S-19: Esquinas Rainforest Lodge: Tropical wet forest. Elev. 800'. Lat. 9°33.131'N, long. 83°48.624'W. This is an excellent ecolodge at the head of the Osa Peninsula near the town of Gamba and Piedras Blancas NP. It has a great variety of wildlife of the southern Pacific lowlands, ranging from Spectacled Owls to antbirds, hummingbirds, Baird's Trogons, and the endemic Black-cheeked Ant-Tanager. www.esquinaslodge.com. Telephone: 506-2741-8001.

S-20: La Cusinga, Oro Verde, Cristal Ballena, Ballena Marine NP: La Cusinga Lodge: Tropical wet forest. Elev. Sea level–400'. Lat. 9°8.50'N, long. 83°42.90'W. This outstanding ecolodge, an ecologically sustainable facility that is sensitive to environmental protection, provides a rustic and attractive setting overlooking the Pacific Ocean. It is possible to see a great variety of rainforest wildlife, including Great Tinamous, Spectacled Owls, manakins, toucans, and oropendolas in the vicinity of the cabins, along the beach, and along the trails. The owners of this property, John Tressemer and his son, Geinier Guzmán, were instrumental in the establishment of the Ballena Marine NP. Address: La Cusinga, S.A., Apdo 41-8000, San Isidro del General, Costa Rica; www.lacusingalodge.com. Telephone: 506-2770-2549. While staying at La Cusinga, it is possible to take a whale- and dolphin-watching excursion to see the migrant Humpback Whales, False Killer Whales, dolphins, and marine birds of the offshore areas in Ballena Marine NP. This national park was created in 1991 by President Oscar Arias after photos taken by the author on a Henderson Birding Tour in 1990 documented the winter calving grounds of Humpback Whales offshore from Caño Island and Punta Uvita. The park includes 425 acres of oceanfront land and 12,750 acres of ocean. For a boat tour to see the whales and dolphins, contact Delfín Tours (506-2743-8169), Ballena Tours (506-2831-1617), or Pelican Tours (506-2743-8047; cabinaslarr@hotmail.com). Other resorts in this vicinity include Cristal Ballena Hotel Resort: Lat. 9°7.441'N, long. 83°41.855'W. www.cristal-ballena.com. Telephone: 506-2786-5354. Whales and Dolphins: A four-star hotel. www.whalesanddolphins.net. Telephone: in USA (toll-free), 1-866-429-3958; in Costa Rica, 506-2743-8150; e-mail: sales@whalesanddolphins.net. Oro Verde Tropical Rainforest Reserve: Tropical wet forest. Lat. 9°12.40'N, long. 83°45.60'W. Elev. 1,000'–2,200'. This private rainforest reserve is about two miles northwest of the village of Uvita along the coast road and two miles east. It does not have lodging available but is a good destination for a day trip to explore the rainforest. It has good trails and an observation tower. Distinctive birds of this site are the Black-crested Coquette, Blue-throated Goldentail, Brown-hooded Parrot, Chestnut-mandibled Toucan, Blue-crowned Motmot, and Crested Owl. www.costarica-birding-oroverde.com. Telephone: 506-2743-8072, 506-2843-8833, 506-2827-3325.

S-21: Hacienda Barú NWR: Tropical wet forest. Elev. Sea level. Lat. 9°15.984'N, long. 83°53.028'W. Playa Dominical and Hacienda Barú NWR are beach areas with migrant shorebirds, like Willets and Whimbrels, and adjacent forest with abundant birdlife. Hacienda Barú Resort; www.haciendabaru.com. Telephone: 506-2787-0003.

S-22: Playa Hermosa: Tropical wet forest. Elev. Sea level. Lat. 9°34.495'N, long. 84°36.673'W. Playa Hermosa and adjacent pastures are noted for wetland birds, many wintering Barn Swallows, and recent records of Southern Lapwings.

CENTRAL PLATEAU

P-1: Sarchí vicinity: Premontane wet forest. Elev. 3,100'. Lat. 10°5.10'N, long. 84°20.80'W. Birds of this urban area include the Blue-gray Tanager, Summer Tanager, Baltimore Oriole, Grayish Saltator, Rufous-tailed Hummingbird, Yellow Warbler, and Clay-colored Thrush.

P-2: Xandari Plantation Resort, Juan Santamaría International Airport–Pavas vicinity, Hotel Alta, and Tobías Bolaños Airport: Premontane moist forest. Elev. 3,200'. Lat. 9°59.60'N, long. 84°08.40'W. The Xandari Plantation Resort has beautifully landscaped grounds with many ornamental flowers and birds. Blue-crowned Motmots, Ferruginous Pygmy-Owls, Tropical Screech-Owls, and even Long-tailed Manakins can be encountered there. The shade coffee plantation on the grounds is one of the only places in Costa Rica to find the Buffy-crowned Wood-Partridge. Address: Xandari Plantation, Apdo 1485-4050, Alajuela, Costa Rica; www.xandari.com. Telephone: 506-2443-2020; e-mail: hotel@xandari.com. Hotel Alta has beautifully landscaped grounds and is an excellent hotel near the airport. www.thealtahotel.com. Telephone: 506-2282-4160; e-mail: info@thealtahotel.com. Hotel Aeropuerto: www.hotelaeropuerto-cr.com; telephone: 2433-7333.

P-3: San José vicinity and downtown: Premontane moist forest (urban). Elev. 3,700'. Lat. 9°56.96'N, long. 84°04.05'W.

P-4: Tres Ríos, Curridibat: Premontane moist forest. Elev. 3,900'. Lat. 9°54.14'N, long. 84°00.37'W.

P-5: Cartago vicinity, Lankester Gardens, and Las Concavas marsh: Premontane moist forest. Elev. 4,700'. Lat. 9°50.20'N, long. 83°53.55'W. The Parque de Expresión in Cartago has ponds with waterbirds like Northern Jacanas. The private Las Concavas marsh and adjacent pastures can be viewed with permission; look for wintering Blue-winged Teal, Killdeer, Eastern Meadowlarks, and Least Grebes.

P-6: Hotel Bougainvillea: Premontane moist forest. Elev. 3,910'. Lat. 9°58.244'N, long. 84°04.965'W. Hotel Bougainvillea in Santo Domingo de Heredia is an excellent hotel for beginning or ending a stay in Costa Rica. The eight acres of gardens are beautifully landscaped and attract a great variety of birds, including Blue-crowned Motmots, Grayish Saltators, Ferruginous Pygmy-Owls, and rare White-eared and Prevost's Ground-Sparrows, which regularly visit the citrus compost site on the east side of the garden. Mailing address: P.O. Box 69-2120, San José, Costa Rica; www.hb.co.cr. Telephone: in USA (toll-free), 1-866-880-5441; in Costa Rica, 506-2244-1414; e-mail: info@hb.co.cr. Motel staff can arrange for rides to the nearby biodiversity park and interpretive center of the National Biodiversity Institute in Santo Domingo de Heredia. www.inbio.ac.cr; www.inbioparque.com. Telephone: INBIO, 506-2507-8100; INBIO Parque, 506-2507-8107.

CARIBBEAN LOWLANDS

C-1: Inocentes–Río Frío region; road from Los Inocentes Ranch

east to lowlands by Santa Cecilia: Tropical moist forest. From lat. 11°02.70'N, long. 85°30.00'W. at Los Inocentes Ranch (now closed to tourism) to lat. 11°03.70'N, long. 85°24.40'W. at Santa Cecilia. This region has wildlife species of the Caribbean lowlands. Owls and Common and Great Potoos can be seen along the road at night with the aid of spotlights.

C-2: Caño Negro NWR and Natural Lodge Caño Negro: Tropical moist forest. Elev. 175'. Lat. 10°54.50'N, long. 84°47.70'W. An exceptional refuge of 24,483 acres, providing habitat for wetland wildlife of the Caribbean lowlands, including waterfowl, wading birds, and rare species like the Nicaraguan Seed-Finch, Nicaraguan Grackle, Agami Heron, and Green Ibis. There is a recent record of nesting Jabiru Storks. The Caño Negro wetlands have been designated one of the most important wetlands in the world. Lodging, boat trips, and tarpon fishing in the nearby lake and channels of the Río Frío can be arranged at Natural Lodge Caño Negro; www.canonegro lodge.com. Telephone: central office, 506-2265-1204, 506-2265-3302, 506-2265-1298; hotel, 506-2471-1000, 506-2471-1426; e-mail: info@canonegrolodge .com.

C-3: Laguna del Lagarto and Ara Ambigua Lodges: Tropical wet forest. Elev. 200'. Lat. 10°41.20'N, long. 84°11.20'W. The area of La Laguna del Lagarto Lodge has wildlife species of the Caribbean lowland rainforest and many wetland species. Address: P.O. Box 995-1007 Centro Colón, San José, Costa Rica; www.lagarto-lodge-costa-rica.com. Telephone: 506-2289-8163; e-mail: info@ lagarto-lodge-costa-rica.com. Ara Ambigua, a rainforest lodge, is north of Puerto Viejo de Sarapiquí; it has a frog garden and opportunities for viewing wildlife along the Río Sarapiquí and forest trails. www.hotelaraambigua .com. Telephone: 506-2766-7101, 506-2766-6401; e-mail: info@ hotelaraambigua.com.

C-4: Tortuga Lodge and Tortuguero NP: Tropical wet forest. Elev. Sea level. Lat. 10°34.36'N, long. 83°31.04'W. An exceptional area of lowland wet forest with great opportunities to view macaws, monkeys, toucans, bats, crocodiles, hummingbirds, and butterflies along the canals and foot trails behind Tortuga Lodge. Address: Costa Rica Expeditions, Apdo 6941-1000, San José, Costa Rica, or Dept. 235, Box 025216, Miami, FL 33102-5216; www.costaricaexpeditions.com. Telephone: 506-2222-0333, 506-2257-0766; e-mail: costaric@ expeditions.co.cr.

C-5: La Selva, Selva Verde, Sueño Azul, Puerto Viejo, and Rancho Gavilán: Tropical wet forest. Elev. 200'. Lat. 10°25.89'N, long. 84°00.27'W. La Selva Biological Field Station is a research station operated by OTS. The author first studied tropical ecology in a course at La Selva in 1969. It is a great destination for observing wildlife of the Caribbean lowlands along well-maintained trails and boardwalks through the forest. There is an excellent opportunity to see owls, motmots, hummingbirds, trogons, antbirds, tinamous, Collared Peccaries, and other rainforest species. Address: Organization for Tropical Studies, Apdo 676-2050, San Pedro de Montes de Oca, San José, Costa Rica; www.threepaths.co.cr. Telephone: in USA, 1-919-684-5774; in Costa Rica, 506-2524-0607; e-mail: edu .travel@ots.ac.cr. Selva Verde Lodge, Sueño Azul Resort, and El Gavilán Lodge are all excellent places to stay while visiting La Selva. It is also possible to stay in cabins at the La Selva OTS facility by contacting the OTS for reservations. Selva Verde Lodge address: Chilamate, Sarapiquí, Costa Rica; www.selevaverde.com. Telephone: in USA (toll-free), 1-800-2451-7111; in Costa Rica, 506-2766-6800; e-mail: selvaver@ racsa.co.cr. El Gavilán Lodge, www.gavilanlodge.com. Telephone: 506-2766-6743, 506-2234-9507; e-mail: gavilan@racsa.co.cr. Sueño Azul Resort is an outstanding lodge a few miles south of La Selva. The grounds and adjacent pastures and river provide excellent birding for such specialties as Fasciated Tiger-Herons, trogons, motmots, guans, Scaled Pigeons, Black-faced Grosbeaks, and Sunbitterns. www.suenoazulresort.com. Telephone: in San José, 506-2253-2020; hotel, 506-2764-1000, 506-2764-1048, 506-2764-1049; e-mail: info@suenoazulresort .com.

C-6: Guacimo, road from La Selva to Guacimo lowland turnoff: Tropical wet forest. Elev. 200'. From lat. 10°25.89'N, long. 84°00.27'W, to lat. 10°13.0'N, long. 83°56.0'W. This is an area of cleared pastureland and scrub, small ponds, and some rivers—good for herons, egrets, anis, and an occasional King Vulture. La Tirimbina BR; www.tirimbina.org. Telephone: 506-2761-1579; e-mail: info@ tirimbina.org.

C-7: Rara Avis: Premontane rainforest. Elev. 2,000'. Lat. 10°17.30'N, long. 84°02.47'W. This 1,500-acre reserve is an excellent place to observe wildlife of the Caribbean lowlands and foothills, including the rare Snowcap hummingbird. Address: Apdo 8105-1000, San José, Costa Rica; www.rara-avis .com. Telephone: 506-2764-1111; e-mail: info@rara-avis.com.

C-8: Rainforest Aerial Tram, Tapir Trail, and Braulio Carrillo NP: Tropical wet forest. Elev. 2,000'. Lat. 10°10.80'N, long. 83°56.60'W. The tram and trails on the property provide excellent places to observe wildlife of middle elevations. Address: Apdo 1959-1002, San José, Costa Rica; www.rfat.com. Telephone: 506-2257-5961; e-mail: info@rfat.com. At Tapir Trail, a private reserve along the highway east of Braulio Carrillo NP, it is possible to see Black-crested Coquettes and Snowcaps.

C-9: Guapiles and Guacimo lowlands: Tropical wet forest. Elev. 900'. Lat. 10°12.85'N, long. 83°47.35'W. The highway from Guapiles east to Limón is excellent for spotting sloths along the highway. Rare Fasciated Tiger-Herons can sometimes be seen on rocks near the Río Roca bridge. East of Guapiles is the famous EARTH University, a tropical sustainable research station for agriculture that also provides rooms for tourists and excellent opportunities to view wildlife on a forest reserve of more than 1,000 acres. www.earth.ac.cr. Telephone: 506-2713-0000.

C-10: Road from Limón to Cahuita: Tropical moist forest. Elev. Sea level. From lat. 9°59.20'N, long. 83°02.00'W, to lat. 9°45.00'N, long. 82°50.20'W. Along this coastal highway it is possible to see sloths, Collared Aracaris, Blue-headed Parrots, and many shorebirds and wading birds in the estuaries that flow into the Caribbean. See safety warning for Cahuita, site C-11.

C-11: Cahuita NP, El Pizote Lodge: Tropical moist forest. Elev. Sea level. Lat. 9°45.00'N long. 82°50.20'W. This national park, encompassing 2,637 acres, was designated for protection of the coral reef there. It is the best example of coral reef in the country, but it has suffered in recent times from chemical pollution and siltation from banana plantations. This is one of the best places in the country to observe sloths, and there are many shorebirds, tanagers, and other wildlife species that can be seen. In late October the Cahuita and Puerto Viejo area is a major passage site for raptors migrating from North America to Panama and South America. Over a million raptors were counted passing through this area during late October 2000 (Jennifer McNicoll-Giancarlo [jmcnicoll@usgs.gov], personal communication). Warning: Violence associated with the local drug culture can make this area unsafe for careless tourists who visit local bars and stray from major hotels and public beaches. El Pizote Lodge is a good lodge there. www.pizotelodge.com. Telephone: 506-2750-0088; e-mail: pizotelg@hotmail.com. Another good lodging facility is Punta Cocles. www.hotelpuntacocles.com. Telephone: 1-888-790-5264; reservations: booking@hotelpuntacocles.com.

C-12: Valle Escondido: Tropical wet forest. Elev. 1,700'. Lat. 10°16.50'N, long. 84°31.80'W. Valle Escondido Lodge is a rainforest resort at the town of San Lorenzo, north of San Ramón. It provides opportunities for birding on 150 acres of primary rainforest and adjacent mixed-forest pastures. Birds are characteristic of the Caribbean lowlands, including Keel-billed Toucans, Red-billed Pigeons, Red-lored Parrots, and White-crowned Parrots. Address: Apdo 452, 1150 La Uruca, Costa Rica; www.costaricareisen.com (go to "hotels"). Telephone: 506-2231-0906.

C-13: Siquirres: Premontane wet forest. Elev. 500'. Lat. 10°6.0'N, 83°30.0'W. This site includes downtown Siquirres, where Tropical Mockingbirds can be seen in the church courtyard. In pasturelands to the northeast it is possible to see Red-breasted Blackbirds. The Costa Rican Amphibian Research Center, eight miles south of Siquirres, has the highest documented diversity of amphibians in Costa Rica. It is also regularly visited by flocks of Great Green Macaws. www.cramphibian.com.

C-14: Río Parismina: Tropical wet forest. Elev: Sea level. Lat. 10°18.388'N, long. 83°21.302'W. The canal from Tortuguero NP to Moin north of Limón provides an excellent opportunity for watching wildlife from a boat. The mouth of the Río Parismina is particularly rich in aquatic birdlife, including Black-necked Stilts, Greater Yellowlegs, Snowy Egrets, Tricolored Herons, Royal Terns, and Willets.

C-15: Jalova: Tropical wet forest. Elev: Sea level. Lat. 10°20.642'N, long. 83°23.935'W. The Jalova ("four corners") field office of Tortuguero NP is along the canal that leads from Tortuguero NP to Moin. Boating along the canal and in the courtyard at the office provides opportunities to see crocodiles, Blue-winged Teals, Golden-hooded Tanagers, Green Honeycreepers, Plumbeous Kites, American Pygmy Kingfishers, and Mangrove Swallows. Even the rare manatee has been seen along the canal between this site and Tortuguero NP.

C-16: Río San Juan: Tropical wet forest. Elev: Sea level. Lat. 10°53.765'N, long. 83°40.826'W. This site is along the Río San Juan near its mouth at the Caribbean, in far northeastern Costa Rica. It is near Río Indio Lodge, which is north across the river in Nicaragua. Wildlife of the area includes the Common Black-Hawk, Mantled Howler Monkey,

Collared Forest-Falcon, crocodile, Three-toed Sloth, Bare-throated Tiger-Heron, Purple-throated Fruitcrow, Strawberry Poison Dart Frog, and the rare White-flanked Antwren. Río Indio Lodge, www.rioindiolodge.com. Telephone: in USA, 1-800-2593-3176; lodge, 506-2296-3338, 506-2296-0095; e-mail: info@rioindiolodge.com.

C-17: Río Sarapiquí: Tropical wet forest. Elev: Sea level. Lat. 10°42.918'N, long. 83°56.314'W. This site is near the mouth of the Río Sarapiquí, where it enters the Río San Juan on the Nicaragua border. Wildlife that can be seen by boat includes crocodiles, Green and Amazon Kingfishers, King Vultures, Green Iguanas, Brazilian Long-nosed Bats, and Mantled Howler Monkeys.

C-18: Tilajari Resort: Tropical moist forest. Elev: 350'. Lat. 10°28.308'N, long. 84°28.099'W. This excellent resort provides good access to the surrounding Arenal volcano area and to the Caño Negro NWR. The grounds provide good birding on forty acres along the Río San Carlos. The lodge also provides access for viewing wildlife on a 600-acre cattle ranch and a 1,000-acre rainforest reserve. Address: Muelle, San Carlos, Costa Rica; www.tilajari.com. Telephone: 506-2462-1212; e-mail: info@tilajari.com.

C-19: Lost Iguana, Arenal, Hanging Bridges: Tropical wet forest. Elev. 1,600'. Lat. 10°29.128'N long. 84°45.316'W. Lost Iguana Resort is a delightful rainforest lodge that provides a spectacular view of the Arenal volcano and excellent wildlife viewing on the 100 acres of habitat on the grounds. Early morning excursions by the lodge can offer sightings of Great Antshrikes, Barred Antshrikes, Dusky Antbirds, Purple-crowned Fairies, and Crested Guans. www.lostiguanaresort.com. Telephone: 506-2479-1555; e-mail: maritzalostiguana@mac.com. Volcano Arenal is the third most active volcano in the world and is part of the 30,000-acre Arenal Volcano NP. Lost Iguana Resort is close to the 250-acre Arenal Hanging Bridges rainforest. The trails and suspended bridges there provide access to great birding in a rainforest setting. www.hangingbridges.com. Telephone: 506-2479-9686. Another excellent lodge in the vicinity is the Arenal Observatory Lodge, owned and operated by one of Costa Rica's most prominent tourism companies, Costa Rica Sun Tours. Ad-dress: Apdo 13411-1000, San José, Costa Rica; www.arenalobservatorylodge.com. Telephone: 506-2692-2070, 506-2290-7011. Another birding location is the Birdhouse B&B. It features tropical gardens, bird feeders, orchids, and trails. Telephone: 506-2694-4428.

HIGHLANDS

H-1: Monteverde Cloud Forest Reserve: Lower montane rainforest. Elev. 4,500'. Lat. 10°19.00'N, long. 84°49.19'W. Monteverde Cloud Forest Reserve (Tropical Science Center) is an excellent example of cloud forest, with Resplendent Quetzals, Three-wattled Bellbirds, Black Guans, and many hummingbirds, including the endemic Coppery-headed Emerald. Telephone: 506-2645-5122; e-mail: montever@sol.racsa.co.cr. Santa Elena Cloud Forest Reserve; www.monteverdeinfo.com; telephone: 506-2645-5390. There are many excellent hotels in the vicinity, including Monteverde Lodge: Apdo 6941-1000, San José, Costa Rica; www.costaricaexpeditions.com; telephone: 506-2645-5057, 506-2257-0766; e-mail: costaric@expeditions.co.cr. Hummingbird Gallery: telephone: 506-2645-5030. Hotel Fonda Vela: Apdo 70060-1000, San José, Costa Rica; www.fondavela.com; telephone: 506-2645-5125; e-mail: info@fondavela.com. Hotel Belmar: Apdo 17-5655, Monteverde, Costa Rica; www.centralamerica.com; telephone: 506-2645-5201; e-mail: belmar@racsa.co.cr.

H-2: Poás NP and Poás Volcano Lodge: Montane rainforest. Elev. 8,200'. Lat. 10°11.45'N, long. 84°13.95'W. Excellent example of montane forest, includes 16,076 acres. It is a good place to see Sooty Thrushes, Yellow-thighed Finches, Large-footed Finches, Magnificent and Volcano Hummingbirds, Slaty Flowerpiercers, and Bare-shanked Screech-Owls. An easy day trip while staying in San José. Poás Volcano Lodge: Lower montane wet forest. Elev. 6,342'. Lat. 10°09.746'N, long. 84°09.516'W. This facility provides convenient lodging near Vara Blanca while visiting Poás NP. Wildlife can be enjoyed in the excellent gardens on the grounds of the lodge, including Black Guans, Violet Sabrewings, Purple-throated Mountain-gems, Ruddy-capped Nightingale-Thrushes, and Bare-shanked Screech-Owls. Address: Apdo 1935-3000, Heredia, Costa Rica; www.poasvolcanolodge.com. Telephone: 506-2482-2194; e-mail: info@poasvolcanolodge.com.

H-3: La Virgen del Socorro: Premontane wet forest. Elev. 2,600', road descending to 2,200'. Lat. 10°15.68'N, long. 84°10.47'W. This road has long been a popular trail for viewing wildlife in the Caribbean foothills. Birds that can be seen in the forest along this road include the White Hawk, Violet-headed Hummingbird, and Black-crested Coquette. From the bridge at the lower end of the road it is possible to see Torrent Tyrannulets and dippers.

H-4: La Paz Waterfall Gardens: Lower montane rainforest. Elev. 4,760'. Lat. 10°12.260'N, long. 84°9.695'W. This outstanding site on the east slope of Volcano Poás has 3.5 kilometers of trails and seventy acres of rainforest, with excellent trails for viewing wildlife of higher elevations, including rare species like the Sooty-faced Finch, dipper, and Azure-hooded Jay. At feeders it is possible to see Crimson-collared Tanagers and Prong-billed Barbets. A hummingbird feeder area hosts local specialties like the Green Thorntail, Brown Violet-ear, White-bellied Mountain-gem, Black-bellied Emerald, and endemic Coppery-headed Emerald. There is also a serpentarium, butterfly observatory, and large aviary. Restaurant and cabins are available. www.waterfallgardens.com. Telephone: 506-2482-2720, ext. 573; 506-2482-2721; e-mail: wgardens@racsa.co.cr.

H-5: El Portico Hotel, San José de la Montaña: Lower montane rainforest. Elev. 5,800'. Lat. 10°05.0'N, long. 84°07.0'W. An excellent location for higher-elevation tanagers, hummingbirds, migrant warblers, and raptors.

H-6: La Ponderosa farm near Turrialba: Premontane wet forest. Elev. 3,760'. Lat. 9°57.31'N, long. 83°42.42'W. Private land. Not accessible for tourism purposes.

H-7: Rancho Naturalista Mountain Lodge: Premontane wet forest. Elev. 3,200'. Lat. 9°49.92'N, long. 83°33.85'W. An exceptional site in the Caribbean foothills, with species of both lowlands and higher elevations, it is one of the best places in the country to see many hummingbirds, including the rare Snowcap, it is the only place where the rare Tawny-chested Flycatcher can be regularly seen. Many birds come to the feeders in the courtyard, and the viewing of hummingbirds at the hummingbird pools in the forest is unique in the country. Excellent naturalist guides and trails. Address: 3428 Hwy 465. Sheridan, AR 72150; www.costaricagateway.com. Telephone: in the USA (toll-free), 1-888-246-8513; reservations, 506-2433-8278; e-mail: crgateway@racsa.co.cr.

H-8: Tapantí NP and Kiri Lodge: Premontane wet forest. Entrance elev. 4,300'. Lat. 9°45.620'N, long. 83°47.038'W. Bridge over the Río Grande de Orosí elev. 5,000'. Lat. 9°42.21'N long. 83°46.93'W. This national park covers 12,577 acres and is a great place to see wildlife of montane forests, like Collared Trogons, Costa Rican Pygmy-Owls, Red-headed and Prong-billed Barbets, Spangle-cheeked Tanagers, dippers, and Azure-hooded Jays. Kiri Lodge telephone: 506-2533-2272. Google "Kiri Lodge, Costa Rica."

H-9: Cerro de la Muerte, San Gerardo de Dota region: Montane rainforest. Four popular birding sites, on the PAH. Kilometer 66 elev., road descending from 8,300' to 7,700'. Lat. 9°40.24'N, long. 83°51.92'W. This site, Finca El Jaular, is a private road on the west side of the PAH that is closed by a large gate. The road can be birded on foot—by permission only—by making arrangements to pay an entrance fee ahead of time (call Savegre Mountain Lodge at the telephone numbers listed for site H-10). Vehicles must be left at the main highway, preferably with someone to watch the vehicle. The land is owned by the Vindas family, who live in the valley at the end of the road. The road descends through excellent primary montane rainforest and is a good place to see quetzals and other highland wildlife. Kilometer 76 elev., 9,400'. Lat. 9°35.68'N, long. 83°48.59'W. Near Los Chespiritos Restaurant 1 is a turnoff to Providencia. Along this road it is possible to see Silvery-throated Jays, Slaty Flowerpiercers, Yellow-thighed Finches, Fiery-throated Hummingbirds, and Black-billed Nightingale-Thrushes. The road is twelve kilometers long, but some of the best birding is in the first two kilometers from the PAH. Kilometer 86 elev., 9,100'. Lat. 9°36.88'N, long. 83°49.07'W. This site is a trail on the west side of the PAH, across the road and a couple hundred feet south of Los Chespiritos Restaurant 2. It is an excellent place to encounter the Timberline Wren, Peg-billed Finch, and high-elevation wildflowers. Kilometer 96 elev. 9,300'. Lat. 9°33.46'N, long. 83°42.67'W. This is west across the PAH from La Georgina restaurant and Villa Mills, at the site of an old highway construction camp where there is shrubby cover that is excellent for Volcano and Scintillant Hummingbirds and Timberline Wrens.

H-10: Savegre: Lower montane rainforest. Elev. 9,400'–7,200'. Lat. 9°32.92'N, long. 83°48.64'W. The turnoff from the PAH at kilometer 80 (at 9,400' elevation) descends for 5.5 kilometers into the valley of San Gerardo de Dota to Savegre Mountain Lodge (Albergue de Montaña Savegre, Cabinas Chacón) along the Río Savegre. This is an excellent area to see Black Guans, Resplendent Quetzals, Long-tailed Silky-Flycatchers, Black-faced Solitaires, Acorn Woodpeckers, Collared Trogons, and resident Red-tailed Hawks. Savegre Mountain Hotel, www.savegre.co.cr. Telephone: 506-2740-1028, 506-2740-1029; in USA (toll-free), 1-800-593-3305.

H-11: Transmission tower site, Cerro de la Muerte: Subalpine rain paramo. Elev. 10,800'. Lat. 9°33.25'N, long. 83°45.16'W. The gravel road leading to the transmission towers from the PAH is approximately at kilometer 90. It is an excellent place to see Volcano Juncos, Peg-billed Finches, resident Red-tailed Hawks, and high-elevation wildflowers.

H-12: Vista del Valle: Elev. 5,650'. Lat. 9°27.78'N, long. 83°42.12'W. Vista del Valle is an excellent spot with both a restaurant and cabins that provide birding along the PAH, at kilometer 119, as the highway begins its descent from Cerro de la Muerte to San Isidro del General. In late January and early February, this is an excellent location to watch Swallow-tailed Kites migrating north from South America. There are feeders that feature the Violet Sabrewing, Red-headed Barbet, Cherrie's Tanager, Flame-colored Tanager, Bay-headed Tanager, and rare White-tailed Emerald. www.vistadelvallecr.com. Telephone: 506-2384-4685.

H-13: Bosque de Paz Ecolodge: Lower montane rainforest. Elev. 4,580'–8,000'. Lat. 10°12.272'N, long. 84°19.032'W. Bosque de Paz is an outstanding lodge on 1,800 acres of montane rainforest on the slope of Poás volcano. Birds of the area include the Resplendent Quetzal, Golden-browed Chlorophonia, Long-tailed Silky-Flycatcher, Purple-throated Mountain-gem, Chestnut-capped Brush-Finch, and American Dipper. Black Guans come to the feeders in the courtyard by day, and tepescuintles, coatis, and agoutis come to the feeders in the evening. Address: P.O. Box 130-1000, San José, Costa Rica; www.bosquedepaz.com. Telephone: 506-2234-6676; e-mail: info@bosquedepaz.com.

H-14: Rincón de la Vieja NP: Premontane wet forest. Elev. 2,400'. Lat. 10°46.363'N, long. 85°20.002'W. This national park covers 34,992 acres, with a good system of trails and an interesting mix of wildlife characteristic of both rainforest and dry forest, including Great Curassows, Crested Guans, Long-tailed Manakins, woodcreepers, White-throated Capuchin monkeys, and Boa Constrictors. There are also some unusual features like hot springs and bubbling mud pits. Telephone: Las Pailas administration office, 506-2661-8139; Guanacaste Conservation Area, 506-2666-5051; e-mail: acg@acguanacaste.ac.cr.

APPENDIX C: COSTA RICAN TRIP PREPARATION CHECKLIST

This trip preparation checklist has been prepared by Carrol and Ethelle Henderson and is based on their experience leading twenty-six birding tours to Costa Rica. The clothing and equipment listed are suggested for a two-week birding or natural history type of tour.

LUGGAGE

One or two pieces of soft, durable, canvas-type bags. Tagged and closed with small TSA padlocks during air travel and storage at hotels. Think light! The less you bring, the easier your travel will be.

CLOTHING

Bring lightweight wash-and-wear clothes you can wash out yourself. Bring detergent double-bagged in self-sealing bags if laundry service is not available.

3–4 sets of field clothes: shirts/blouses; pants, shorts, or jeans; and one long-sleeved shirt.
Socks (4–5 pairs)
Underwear (4–5 pairs)
Handkerchiefs or tissues
Belt
Sweatshirt/sweater/light jacket
Hat or cap
Sleepwear
One pair walking shoes; one pair tennis or hiking shoes
Rain poncho or raincoat (lightweight)
Swimsuit and beach thongs

TOILETRY ITEMS

Pack of Wet Ones or similar towelettes
Deodorant
Shaving cream
Toothbrush
Toothpaste/dental floss
Shampoo, without citronella base
Comb/hairbrush
Razor/shaver (electric current is 110 AC, but some outlets don't take wide prong; bring adapter plug)

PHOTO AND OPTIC EQUIPMENT

Camera with flash unit or video camera
Binoculars
Camera bag
Extra batteries for camera and flash unit and/or battery recharger
Lenses and filters
Lens tissue
Memory chips/film: 4 gigs—average interest in photos; 8 gigs—moderate interest; 12 gigs—enthusiastic. (Bring more than you think you will need. Film and memory chips can be hard to find and very expensive to buy on the road.)

OTHER EQUIPMENT

Fingernail clippers
Sunglasses
Suntan lotion or sunscreen (at least SPF 30)
Chapstick
Insect repellent (up to 30 percent DEET)
Aspirin
Imodium or Lomotil
Q-Tips
Notebook and pens
Small flashlight
Field guides
Knapsack/daypack/fannypack
Spending money, at least $400 in clean, undamaged bills (U.S. currency and credit cards are accepted in most hotels and larger stores. Call your credit card company before you go to let them know you will be in Costa Rica. You can change some currency to Costa Rican colones at most hotels. Use colones in small towns.)
Prescriptions for personal medication, including original containers.
Travel alarm
Passport, plus photocopy packed separately from passport
Tip for naturalist guide (about $10–$12 per day)
Tip for driver (about $6–$8 per day)
Earplugs for sleeping near noisy highways or near loud surf
Umbrella (compact)

OTHER OPTIONAL ITEMS

Hunting or fishing vest for gear
Spare camera
Mending kit
Water bottle
20' cord for indoor clothesline

MAP

There is one exceptional map for Costa Rica that shows topographical features in great detail. Called a "tactical pilotage map," it is published by the U.S. Department of Defense. These maps are available for all regions of Latin America in a scale of 1:500,000. The map for Costa Rica is TPC K-25C. It can be ordered from the Latitudes Map and Travel Store in Minneapolis, Minnesota (www.latitudesmapstore.net), or from the NOAA Distribution Branch (N/CG33, National Ocean Service, Riverdale, MD 20737).

APPENDIX D: TRAVEL TIPS FOR A SUCCESSFUL WILDLIFE VIEWING TRIP IN COSTA RICA

1. Begin trip planning at least six months prior to your trip. The best lodges fill early during their high season from January to March. This is the dry season, which is typically the best time to visit Costa Rica. The first half of July can also be a good time to go.

2. Decide if you wish to travel independently or participate in a birding tour. A well-organized birding tour with a good guide and outfitter company will take care of logistics, driving, lodging arrangements, meals, and safety considerations. You will typically see two to four times more birds on a guided tour than if you travel by yourself.

3. Be aware that there are several levels of intensity for birding tour groups. Some groups are determined to see the maximum number of birds in the time available, around 400-plus species in two weeks. The pace is intense and is focused only on birds. Other birding groups are moderately paced. You may see about 300–350 bird species in two weeks with a group that is still focused primarily on birds but takes time to enjoy a broader spectrum of the flora and fauna, like butterflies, flowers, and culture. General natural history groups are more passive, walk less, and are broadly interested in nature; you will see perhaps 100–125 species in two weeks. Get references and contact former clients to make sure you sign up for a group that matches your expectations, interests, and preferred level of physical activity.

4. Traveling by yourself can be cheaper, but you need to deal with lodging, meals, travel arrangements, and the Spanish language. If traveling by yourself, visit lodges that have naturalist birding guides, or lodges where you can hire local birding guides for day trips. Otherwise, hire a birding guide to accompany you on your entire trip.

5. When you finish eating at a restaurant, always check your tables and chair backs for binoculars, cameras, sunglasses, daypacks, and other items for yourself and for other members of your party.

6. While birding in Costa Rica, share the experience. After you have spotted a bird, help others in the group find it if they can't see it. If you encounter other birders or Costa Rican families while birding, let them take a peek through your spotting scope or binoculars if they have no optics.

7. When traveling in a bus, sit in a different seat every day to give everyone equal access to the best seats.

8. When birding along a narrow trail, switch positions with others every fifteen to twenty minutes to avoid dominating the best positions behind the guide.

9. When organizing your itinerary, try to include at least three of Costa Rica's biological zones (for example, Guanacaste, highlands, and Caribbean slope).

10. Bring any trip problems or complaints about your tour to the attention of your guide or tour leader in a discreet manner if you feel there is a problem in protocol, behavior, or group etiquette that needs to be addressed. Do not wait until after the trip to complain.

About the Author

Carrol L. Henderson, a native of Zearing, Iowa, received a bachelor of science degree in zoology from Iowa State University in 1968 and a master of forest resources degree in ecology from the University of Georgia in 1970. He did his graduate studies on the fish and wildlife of Costa Rica through the Organization for Tropical Studies and the University of Costa Rica.

Henderson joined the Minnesota Department of Natural Resources (DNR) in 1974 as assistant manager of the Lac qui Parle Wildlife Management Area near Milan. In 1977 he became supervisor of the DNR's newly created Nongame Wildlife Program, and he continues in that role to the present. During the past thirty-three years, Henderson developed a statewide program for the conservation of Minnesota's nongame wildlife and has planned and developed projects to help bring back bluebirds, Bald Eagles, Peregrine Falcons, River Otters, and Trumpeter Swans.

Henderson received the national Chevron Conservation Award in 1990, the 1992 Chuck Yeager Conservation Award from the National Fish and Wildlife Foundation, the 1993 Minnesota Award from the Minnesota Chapter of The Wildlife Society, and the 1994 Thomas Sadler Roberts Memorial Award from the Minnesota Ornithologists' Union.

His writings include *Woodworking for Wildlife; Landscaping for Wildlife; Wild about Birds: The DNR Bird Feeding Guide;* and coauthorship of *The Traveler's Guide to Wildlife in Minnesota* and *Lakescaping for Wildlife and Water Quality.* He also wrote the first edition of the *Field Guide to the Wildlife of Costa Rica* in 2002; *Oology and Ralph's Talking Eggs* in 2007; and *Birds in Flight: The Art and Science of How Birds Fly* in 2008. He is a regular contributor of feature stories in *Birder's World* and *Seasons* magazines.

The author with his wife, Ethelle, in Costa Rica, 2005.

An avid wildlife photographer, Henderson has taken most of the photos in his books and was the primary photographer for the 1995 book *Galápagos Islands: Wonders of the World.* His bird photography has been featured in the *New York Times, World Book Encyclopedia of Science, Audubon* magazine, and Discovery Online. He received seven national bird photography awards from *Wild Bird* magazine between 1995 and 1998.

Henderson and his wife, Ethelle, developed their expertise in tropical wildlife by leading forty-four birding tours to Latin America, Africa, and New Zealand since 1987. This includes twenty-six trips to Costa Rica and additional trips to Panama, Belize, Nicaragua, Trinidad, Tobago, Venezuela, Bolivia, Chile, Ecuador, Brazil, Argentina, Peru, the Galápagos Islands, Kenya, Tanzania, and New Zealand.

INDEX

Spanish names for animals are set in **boldface text**.